SpringerBriefs in Energy

SpringerBriefs in Energy presents concise summaries of cutting-edge research and practical applications in all aspects of Energy. Featuring compact volumes of 50 to 125 pages, the series covers a range of content from professional to academic. Typical topics might include:

- A snapshot of a hot or emerging topic
- A contextual literature review
- A timely report of state-of-the art analytical techniques
- An in-depth case study
- A presentation of core concepts that students must understand in order to make independent contributions.

Briefs allow authors to present their ideas and readers to absorb them with minimal time investment.

Briefs will be published as part of Springer's eBook collection, with millions of users worldwide. In addition, Briefs will be available for individual print and electronic purchase. Briefs are characterized by fast, global electronic dissemination, standard publishing contracts, easy-to-use manuscript preparation and formatting guidelines, and expedited production schedules. We aim for publication 8–12 weeks after acceptance.

Both solicited and unsolicited manuscripts are considered for publication in this series. Briefs can also arise from the scale up of a planned chapter. Instead of simply contributing to an edited volume, the author gets an authored book with the space necessary to provide more data, fundamentals and background on the subject, methodology, future outlook, etc.

SpringerBriefs in Energy contains a distinct subseries focusing on Energy Analysis and edited by Charles Hall, State University of New York. Books for this subseries will emphasize quantitative accounting of energy use and availability, including the potential and limitations of new technologies in terms of energy returned on energy invested. The second distinct subseries connected to SpringerBriefs in Energy, entitled Computational Modeling of Energy Systems, is edited by Thomas Nagel, and Haibing Shao, Helmholtz Centre for Environmental Research - UFZ, Leipzig, Germany. This sub-series publishes titles focusing on the role that computer-aided engineering (CAE) plays in advancing various engineering sectors, particularly in the context of transforming energy systems towards renewable sources, decentralized landscapes, and smart grids.

All Springer brief titles should undergo standard single-blind peer-review to ensure high scientific quality by at least two experts in the field.

Jinsen Hu • Yuning Zhang • Yongpan Cheng

Multiphase Flow with Bubbles

 Springer

Jinsen Hu
Key Laboratory of Power Station Energy
Transfer Conversion and System, Ministry
of Education, North China Electric Power
University, Beijing, Beijing, China

Yuning Zhang
Key Laboratory of Power Station Energy
Transfer Conversion and System, Ministry
of Education, North China Electric Power
University, Beijing, Beijing, China

Yongpan Cheng
Key Laboratory of Power Station Energy
Transfer Conversion and System, Ministry
of Education, North China Electric Power
University, Beijing, Beijing, China

ISSN 2191-5520   ISSN 2191-5539   (electronic)
SpringerBriefs in Energy
ISBN 978-3-031-99215-5   ISBN 978-3-031-99216-2   (eBook)
https://doi.org/10.1007/978-3-031-99216-2

Preface

Multiphase flow with bubbles represents a complex field at the intersection of fluid dynamics, acoustics, and mechanical engineering. With wide-ranging applications in biomedical engineering, acoustic chemistry, surface cleaning, material processing, and hydraulic machinery, the study of bubble behavior in multiphase systems has become a critical component of modern fluid mechanics. This book presents a comprehensive exploration of bubble dynamics under acoustic excitation and the intricate interplay between bubbles, particles, and sound waves in various multiphase environments.

The journey begins with an in-depth analysis of bubble dynamics under acoustic excitation, where we derive oscillation equations, explore mass transfer effects, and investigate energy dissipation mechanisms. The complexities of dual-frequency acoustic fields and their impact on bubble behavior are also examined. We then transition to the propagation of sound waves in vapor/gas/liquid multiphase flows, presenting predictive models and analyzing wave speed characteristics influenced by factors, such as vapor fraction and bubble radius. The exploration culminates in the study of cavitation in vapor/liquid/solid systems, where bubble collapse, jet formation, shock wave propagation, and their synergistic effects on hydraulic machinery erosion are meticulously detailed.

Designed for researchers, graduate students, and professionals, this work aims to consolidate current knowledge, highlight unresolved challenges, and inspire further investigation. We hope this work will empower readers to push the boundaries of knowledge in multiphase flow, contributing to innovations in engineering and technology.

Beijing, Beijing, China
May 2025
Jinsen Hu
Yuning Zhang
Yongpan Cheng

Acknowledgment

This book was financially supported by the National Natural Science Foundation of China (Project No. 51976056).

Competing Interests The authors have no competing interests to declare that are relevant to the content of this manuscript.

Contents

About the Authors

Jinsen Hu received his Ph.D. from North China Electric Power University in June 2024. His research interests include cavitation and bubble dynamics, with a particular focus on experimental and numerical investigations of bubble collapse, micro-jets, and shock waves. He has contributed to the publication of ten journal papers in these areas, in which he explored the splitting behavior of cavitation bubbles near particles and elucidated the formation mechanisms and evolution of jets and shock waves.

Yuning Zhang is currently a Professor at North China Electric Power University. He primarily focuses on the research in cavitation and bubble dynamics. He has published two monographs with Springer Press and over 90 papers (ten highly cited papers and one hot paper) in journals, such as *Nature Communications*, *Physics of Fluids*, and *Energy*. He was ranked by Stanford University as one of the "top 2% of scientists in the world" in 2023. In addition, he serves as an Associate Editor of IET Renewable Power Generation and as an editorial board member of six international/national journals, such as *Journal of Hydrodynamics*. He also serves as Chairman of International Cavitation Forum 2016 and has been a member of the organizing committee for the WIMRC International Symposium of Cavitation (UK) and many other international conferences. Additionally, he has given six invited talks at international conferences. He has received many awards from the Society of Energy and Electric Power in China.

Yongpan Cheng is mainly engaged in research on the dynamics of bubble and droplet. He is Marie Curie Fellow in the European Union. So far he has published more than 50 journal papers in journals, such as *Langmuir*, *Physics of Fluids*, and *International Journal of Heat and Mass Transfer*, with over 2000 citations. He serves as the editorial board member for the *International Journal of Hydromechatronics*. He has co-organized several international conferences, such as International Heat Transfer Symposium (2014 and 2016) and has served as a guest editor for several journals, such as *Applied Thermal Engineering* and *Industrial & Engineering Chemistry Research*. He has delivered more than ten invited talks at international conferences.

Chapter 1
Introduction

1.1 Research Background

Multiphase flow with bubbles represents a fundamental and interdisciplinary domain within fluid dynamics, characterized by the complex interactions of gas, liquid, and sometimes solid phases. The presence of bubbles in a liquid medium can significantly influence flow properties, energy transfer, and reaction kinetics, making this field pivotal to both scientific inquiry and practical applications. This book explores bubble dynamics in multiphase flows, with a particular focus on their behavior under acoustic excitation and the cavitation phenomena in vapor/liquid/solid systems.

The relevance of this research extends across a diverse array of fields. In hydraulic machinery, such as pumps [1–3], turbines [4–6], and propellers [7, 8], cavitation—the formation and collapse of bubbles—poses a persistent challenge by causing material erosion and diminishing operational efficiency [9]. Specifically, the collapse of these bubbles generates high-speed jets and shock waves that repeatedly impact the surfaces of components like impellers and propellers, leading to pitting, erosion, and material fatigue over time. Insights into bubble dynamics enable the development of strategies to mitigate these effects, enhancing equipment durability and performance. In biomedical engineering, acoustically excited bubbles play a transformative role in applications. Specifically, in ultrasound imaging, bubbles enhance the scattering of sound waves, acting as contrast agents to improve image resolution and diagnostic precision [10, 11]. For targeted drug delivery, the growth and subsequent collapse of bubbles produce high-speed jets and shock waves that penetrate or permeabilize cell membranes, enabling precise delivery of drugs to specific tissues [12, 13]. In lithotripsy, the shock waves generated by bubble collapse fragment kidney stones non-invasively, offering an effective alternative to surgery [14, 15]. In the field of acoustic chemistry, bubble collapse induced by an

J. Hu et al., *Multiphase Flow with Bubbles*, SpringerBriefs in Energy,
https://doi.org/10.1007/978-3-031-99216-2_1

acoustic field generates localized hot spots with very high temperatures and pressures [16–18], which promote chemical reactions.

Further applications include surface cleaning, where bubble dynamics enhance the removal of contaminants [19, 20]. In material processing, cavitation is exploited for precision machining and surface modification, while in alloy strengthening, controlled bubble collapse improves material hardness and wear resistance [21–23]. Additionally, environmental engineering employs bubbles in processes like water purification [24, 25], and mineral processing relies on bubble-particle interactions for effective flotation techniques in resource extraction [26, 27].

1.2 Research Status

The study of bubble motion in liquids underpins bubble dynamics research. Lord Rayleigh [28] pioneered this field with the Rayleigh equation, which describes the radial motion of a spherical bubble in an infinite, incompressible, inviscid, and quiescent liquid. By accounting for liquid viscosity, Plesset [29], Poritsky [30], and Keller and Kolodner [31] each modified Rayleigh equation, giving rise to the Rayleigh–Plesset equation [32]. This model, however, is limited to scenarios with small external driving forces and bubble interface velocities. In cases of vigorous oscillations, the velocity of motion at the bubble interface is no longer a small quantity compared to the sound velocity in the liquid, and the assumption of incompressibility of the liquid no longer holds. To address this, Keller and Miksis integrated liquid compressibility, yielding the Keller–Miksis equation [33].

Under acoustic excitation, bubble oscillations exhibit pronounced nonlinear behavior. Lauterborn [34] investigated primary resonance, harmonic resonance, sub-harmonic and super-harmonic resonances of bubbles under single-frequency acoustic driving, finding that increased ultrasound pressure induces successive regimes of radial, surface, and chaotic oscillations on rigid-surface bubbles. Zhang et al. [35] further examined how initial bubble radius, ultrasound frequency, and pressure amplitude affect interface stability, reporting that pressure amplitude strongly influences spherical stability but has little effect on static stability. Under dual-frequency excitation, bubbles display even more complex dynamics. Zhang et al. [36] observed combination resonances that amplify oscillation amplitude and scattering signals. They also demonstrated that dual-frequency driving lowers the cavitation threshold while producing larger oscillation radii and collapse pressures compared to single-frequency excitation [37]. Guédra et al. [38] reported that dual-frequency excitation nearly doubles the total acoustic power response, enhancing cavitation. When the second frequency is lower, dual-frequency ultrasound prolongs collapse time and increases collapse ratio, whereas a higher second frequency increases the number of oscillation cycles [39].

Mass transfer and energy dissipation are critical in acoustically driven bubbles. In terms of mass transfer, Blake [40] derived a diffusion solution for small-amplitude sinusoidal oscillations, considering only area variations. Epstein and Plesset [41] developed a quasi-steady diffusion model by omitting convective terms. Hsieh and Plesset [42] added convection via a Taylor-series expansion for the moving boundary, though still under small sine-wave constraints. Eller and Flynn [43] used boundary-layer analysis in a Lagrangian frame to model non-sinusoidal oscillations, removing the small-amplitude limit and introducing a threshold-theory framework. Crum [44] expanded this beyond isothermal assumptions. Zhang [45] refined the model by incorporating energy dissipation, inhomogeneous internal pressure, surface tension, and liquid compressibility. Energy dissipation occurs via viscous, thermal, and acoustic losses. Viscous dissipation dominates in small vapor or gas bubbles, thermal dissipation in intermediate-sized bubbles, and acoustic dissipation in large bubbles [46].

When bubbles collapse near solid boundaries, such as rigid walls or particles, they generate microjets, shock waves, and fragmentation. Jet and shock behavior near rigid surfaces is particularly significant. Bußmann et al. [47] studied the relationship between jet behavior and stand-off distance (bubble–wall distance nondimensionalized by maximum bubble radius), identifying three jet regimes: needle jet, transitional jet, and regular jet, as distance increases. Needle jets, characterized by narrow, high-velocity, have been confirmed by Lechner et al.'s simulations [48, 49] and Reuter et al.'s experiments [50]. During jet development, shock waves arise inside the bubble, as the jet penetrates the bubble interface, and upon wall impact [51–53]. Lai et al. [54] focused on the initial shock wave from cavitation bubble, discussing its attenuation characteristics, propagation speed, and intensity. For a bubble near particles, Yu et al. [55] investigated bubble dynamics experimentally, theoretically, and numerically, with emphasis on bubble splitting and the mechanisms of jet and shock-wave formation and evolution.

1.3 Description of the Book

This book provides an in-depth study of multiphase flow with bubbles, covering bubble dynamics under acoustic excitation, the propagation of sound waves in vapor/gas/liquid multiphase flow, and vapor/liquid/solid multiphase flow. For bubble dynamics under acoustic excitation, the oscillation equations of a single bubble in liquids, including first-and second-order equations is first introduced. Then, the mass transfer effect of the bubble wall, the dissipation mechanism of bubble energy, and the bubble dynamics under dual-frequency acoustic fields are analyzed. Regarding the propagation of sound waves in vapor/gas/liquid multiphase flow, a prediction model of wave speed is described, the solution process and the characteristics of wave propagation are elaborated in detail, and the influences of vapor fraction, bubble radius, and void fraction on the critical frequency and constant wave speed are deeply discussed. In the part of vapor/liquid/solid multiphase flow, the bubble collapse process is introduced in detail, the jet phenomena including single jet, multi-jet, and needle jet are analyzed, as well as the shock wave phenomena generated during the nucleation, splitting, and collapse processes of cavitation bubble. Additionally, the synergistic erosion of cavitation bubble and sand particles on hydraulic machinery is introduced. This book is targeted at academic researchers and graduate students in the field of fluid dynamics, aiming to consolidate the basic theories, physical mechanisms, and latest advancements in the field of multiphase flow with bubbles.

References

1. Binama M, Muhirwa A, Bisengimana E. Cavitation effects in centrifugal pumps-a review. Int J Eng Res Appl. 2016;6:52–63.
2. Kan K, Binama M, Chen H, et al. Pump as turbine cavitation performance for both conventional and reverse operating modes: a review. Renew Sust Energ Rev. 2022;168:112786.
3. Tao R, Xiao R, Wang F, et al. Cavitation behavior study in the pump mode of a reversible pump-turbine. Renew Energy. 2018;125:655–67.
4. Kumar P, Saini RP. Study of cavitation in hydro turbines—a review. Renew Sust Energ Rev. 2010;14(1):374–83.
5. Singh R, Tiwari SK, Mishra SK. Cavitation erosion in hydraulic turbine components and mitigation by coatings: current status and future needs. J Mater Eng Perform. 2012;21:1539–51.
6. Adhikari RC, Vaz J, Wood D. Cavitation inception in crossflow hydro turbines. Energies. 2016;9(4):237.
7. Peters A, Lantermann U, el Moctar O. Numerical prediction of cavitation erosion on a ship propeller in model-and full-scale. Wear. 2018;408:1–12.
8. Taseidifar M, Antony J, Pashley RM. Prevention of cavitation in propellers. Substantia. 2020;4(2 Suppl):109–17.
9. Li SC. Cavitation of hydraulic machinery. World Scientific; 2000.
10. Helfield B, Zou Y, Matsuura N. Acoustically-stimulated nanobubbles: opportunities in medical ultrasound imaging and therapy. Front Phys. 2021;9:654374.
11. van Wamel A, Mühlenpfordt M, Hansen R, et al. Ultrafast microscopy imaging of acoustic cluster therapy bubbles: activation and oscillation. Ultrasound Med Biol. 2022;48(9):1840–57.
12. Jeong J, Jang D, Kim D, et al. Acoustic bubble-based drug manipulation: carrying, releasing and penetrating for targeted drug delivery using an electromagnetically actuated microrobot. Sensors Actuators A Phys. 2020;306:111973.
13. Coussios CC, Roy RA. Applications of acoustics and cavitation to noninvasive therapy and drug delivery. Annu Rev Fluid Mech. 2008;40(1):395–420.
14. Xiang G, Chen J, Ho D, et al. Shock waves generated by toroidal bubble collapse are imperative for kidney stone dusting during holmium: YAG laser lithotripsy. Ultrason Sonochem. 2023;101:106649.
15. Ghorbani M, Oral O, Ekici S, et al. Review on lithotripsy and cavitation in urinary stone therapy. IEEE Rev Biomed Eng. 2016;9:264–83.
16. Mettin R, Cairós C, Troia A. Sonochemistry and bubble dynamics. Ultrason Sonochem. 2015;25:24–30.
17. Wood RJ, Lee J, Bussemaker MJ. A parametric review of sonochemistry: control and augmentation of sonochemical activity in aqueous solutions. Ultrason Sonochem. 2017;38:351–70.
18. Luo J, Fang Z, Smith RL, et al. Fundamentals of acoustic cavitation in sonochemistry. In: Fang Z, Smith JL, Qi X, editors. Production of biofuels and chemicals with ultrasound, biofuels and biorefineries. Dordrecht: Springer; 2015.
19. Reuter F, Lauterborn S, Mettin R, et al. Membrane cleaning with ultrasonically driven bubbles. Ultrason Sonochem. 2017;37:542–60.
20. Chahine GL, Kapahi A, Choi JK, et al. Modeling of surface cleaning by cavitation bubble dynamics and collapse. Ultrason Sonochem. 2016;29:528–49.

21. Yoshimura T, Tanaka K, Yoshinaga N. Nano-level material processing by multifunction cavitation. Nanosci Nanotechnol Asia. 2018;8(1):41–54.
22. Soyama H, Korsunsky AM. A critical comparative review of cavitation peening and other surface peening methods. J Mater Process Technol. 2022;305:117586.
23. Galloni MG, Fabbrizio V, Giannantonio R, et al. Applications and applicability of the cavitation technology. Curr Opin Chem Eng. 2025;48:101129.
24. Fetyan NAH, Salem Attia TM. Water purification using ultrasound waves: application and challenges. Arab J Basic Appl Sci. 2020;27(1):194–207.
25. Bandala ER, Rodriguez-Narvaez OM. On the nature of hydrodynamic cavitation process and its application for the removal of water pollutants. Air Soil Water Res. 2019;12:0–6.
26. Zhou ZA, Xu Z, Finch JA, et al. On the role of cavitation in particle collection in flotation–a critical review II. Minerals Eng. 2009;22(5):419–33.
27. Chen Y, Truong VNT, Bu X, et al. A review of effects and applications of ultrasound in mineral flotation. Ultrason Sonochem. 2020;60:104739.
28. Rayleigh L. On the pressure developed in a liquid during the collapse of a spherical cavity. Philos Mag. 1917;34(200):94–8.
29. Plesset MS. The dynamics of cavitation bubbles. J Appl Mech. 1949;16:277–82.
30. Poritsky H. The collapse or growth of a spherical bubble or cavity in a viscous fluid. America J Appl Mech Trans ASME. 1951;18(3):332–3.
31. Keller JB, Kolodner II. Damping of underwater explosion bubble oscillations. J Appl Phys. 1956;27(10):1152–61.
32. Plesset MS, Prosperetti A. Bubble dynamics and cavitation. Annu Rev Fluid Mech. 1977;9(1):145–85.
33. Keller JB, Miksis M. Bubble oscillations of large amplitude. J Acoust Soc Am. 1980;68(2):628–33.
34. Lauterborn W. Numerical investigation of nonlinear oscillations of gas bubbles in liquids. J Acoust Soc Am. 1976;59(2):283–93.
35. Zhang Y, Gao Y, Du X. Stability mechanisms of oscillating vapor bubbles in acoustic fields. Ultrason Sonochem. 2018;40:808–14.
36. Zhang Y, Zhang Y, Li S. Combination and simultaneous resonances of gas bubbles oscillating in liquids under dual-frequency acoustic excitation. Ultrason Sonochem. 2017;35:431–9.
37. Ye L, Zhu X, Liu Y. Numerical study on dual-frequency ultrasonic enhancing cavitationeffect based on bubble dynamic evolution. Ultrason Sonochem. 2019;59:104744.
38. Guédra M, Inserra C, Gilles B, et al. Numerical investigations of single bubble oscillations generated by a dual frequency excitation. J Phys Conf Ser. 2015;656:12019.
39. Kerboua K, Hamdaoui O. Numerical investigation of the effect of dual frequency sonication on stable bubble dynamics. Ultrason Sonochem. 2018;49:325–32.
40. Blake FG. The onset of cavitation in liquids. I. Cavitation threshold sound pressures in water as a function of temperature and hydrostatic pressure, vol. 12. Harvard University Acoustic Res Lab Tech Memorandum; 1949. p. 1–52.
41. Epstein PS, Plesset MS. On the stability of gas bubbles in liquid-gas solutions. J Chem Phys. 1950;18(11):1505–9.
42. Hsich DY, Plesset MS. Theory of rectified diffusion of mass into gas bubbles. J Acoust Soc Am. 1961;33(2):206–15.
43. Eller A, Flynn HG. Rectified diffusion during nonlinear pulsations of cavitation bubbles. J Acoust Soc Am. 1965;37(3):493–503.
44. Crum LA, Hansen GM. Generalized equations for rectified diffusion. J Acoust Soc Am. 1982;72(5):1586–92.
45. Zhang Y, Li S. A general approach for rectified mass diffusion of gas bubbles in liquids under acoustic excitation. J Heat Transf. 2014;136(4):042001.
46. Zhang Y. Heat transfer across interfaces of oscillating gas bubbles in liquids under acoustic excitation. Int Commun Heat Mass Transfer. 2013;43:1–7.

47. Bußmann A, Riahi F, Gökce B, et al. Investigation of cavitation bubble dynamics near a solid wall by high-resolution numerical simulation. Phys Fluids. 2023;35(1):016115.
48. Lechner C, Lauterborn W, Koch M, et al. Fast, thin jets from bubbles expanding and collapsing in extreme vicinity to a solid boundary: a numerical study. Physical Rev Fluids. 2019;4(2):021601.
49. Lechner C, Lauterborn W, Koch M, et al. Jet formation from bubbles near a solid boundary in a compressible liquid: numerical study of distance dependence. Physical Rev Fluids. 2020;5(9):093604.
50. Reuter F, Ohl CD. Supersonic needle-jet generation with single cavitation bubbles. Appl Phys Lett. 2021;118(13):134103.
51. Yang X, Liu C, Wan D, et al. Numerical study of the shock wave and pressure induced by single bubble collapse near planar solid wall. Phys Fluids. 2021;33(7):073311.
52. Zhang M, Chang Q, Ma X, et al. Physical investigation of the counterjet dynamics during the bubble rebound. Ultrason Sonochem. 2019;58:104706.
53. Tian L, Zhang YX, Yin JY, et al. Study on the liquid jet and shock wave produced by a near-wall cavitation bubble containing a small amount of non-condensable gas. Int Commun Heat Mass Transfer. 2023;145:106815.
54. Lai G, Geng S, Zheng H, et al. Early dynamics of a laser-induced underwater shock wave. J Fluids Eng. 2022;144(1):011501.
55. Yu J, Wang X, Shen J, et al. Physics of cavitation near particles. J Hydrodyn. 2024;36(1):102–18.

Chapter 2
Bubble Dynamics Under Acoustic Excitation

In this chapter, the free oscillatory characteristics of a single bubble are first presented, including the derivation of the bubble wall motion equation with the second-order Mach number, and its comparison with the oscillation properties of the bubble calculated by the first-order equation. Then, the mass transfer effect of the bubble wall is analyzed. Finally, the energy dissipation mechanism of bubbles under acoustic excitation and the bubble dynamics under dual-frequency acoustic excitation are discussed.

2.1 Oscillation of a Single Bubble

2.1.1 Basic Equations

Assume that during bubble oscillation, the bubble remains spherical and its center does not displace. The liquid surrounding the bubble is compressible and exhibits irrotational flow, with the gas pressure inside the bubble being uniform. Mass transfer at the bubble wall, liquid viscosity, and body forces are neglected. In spherical coordinates, the continuity equation and momentum equation for the liquid are given as follows [1].

$$\frac{\partial \rho_1}{\partial t} + \frac{1}{r^2} \frac{\partial \left(r^2 \rho_1 u \right)}{\partial r} = 0 \tag{2.1}$$

$$\frac{\partial u}{\partial t} + u \frac{\partial u}{\partial r} + \frac{1}{\rho_1} \frac{\partial p}{\partial r} = 0 \tag{2.2}$$

© The Author(s), under exclusive license to Springer Nature
Switzerland AG 2025
J. Hu et al., *Multiphase Flow with Bubbles*, SpringerBriefs in Energy,
https://doi.org/10.1007/978-3-031-99216-2_2

where ρ_l and $u(r,t)$ represent the density and velocity of the liquid, respectively. r is the radial distance in the spherical coordinate system, and t is time.

The modified Tait equation of state for the liquid is [2].

$$\frac{p+B}{p_\infty+B} = \left(\frac{\rho_l}{\rho_{l,\infty}}\right)^{n_l} \tag{2.3}$$

where the liquid constants B and n_l are experimentally determined as 3049.13 bars and 7.15, respectively. p is the pressure, p_∞ is the pressure at infinity.

The speed of acoustic c_l and enthalpy h in the liquid are defined as [1].

$$c_l^2 = \frac{dp}{d\rho_l} \tag{2.4}$$

$$h = \int_{p_\infty}^{p} \frac{dp}{\rho_l} \tag{2.5}$$

where $c_{l,\infty}^2 = n_l(p_\infty+B)/\rho_\infty$ is the speed of acoustic at infinity.

From Eq. (2.3), the explicit expressions for the speed of acoustic c_l and enthalpy h are [1].

$$c_l^2 = \frac{n_l(p+B)}{\rho_l} = c_{l,\infty}^2 + (n_l-1)/h \tag{2.6}$$

$$h = \frac{c_l^2 - c_{l,\infty}^2}{n_l-1} = \frac{c_{l,\infty}^2}{n_l-1}\left[\left(\frac{p+B}{p_\infty+B}\right)^{\frac{n_l-1}{n_l}} - 1\right] \tag{2.7}$$

For irrotational flow the velocity satisfies [1].

$$u = \frac{\partial\varphi}{\partial r} \tag{2.8}$$

where $\varphi(r,t)$ is the velocity potential. Substituting Eqs. (2.4) and (2.5) into Eq. (2.1) yield [1].

$$\nabla^2\varphi + \frac{1}{c_l^2}\left(\frac{\partial h}{\partial t} + \frac{\partial\varphi}{\partial t}\frac{\partial h}{\partial r}\right) = 0 \tag{2.9}$$

Substituting Eqs. (2.4) and (2.5) into Eq. (2.2) and integrating once with respect to r gives the Bernoulli integral [1].

$$\frac{\partial \varphi}{\partial t} + \frac{1}{2}\left(\frac{\partial \varphi}{\partial r}\right)^2 + h = 0 \tag{2.10}$$

Combining Eqs. (2.7), (2.9), and (2.10), the relationship between φ, r and t is [1].

$$\left[1 - \frac{n_1 - 1}{c_\infty^2}\left(\frac{\partial \varphi}{\partial t} + \frac{1}{2}\left(\frac{\partial \varphi}{\partial r}\right)^2\right)\right]\nabla^2 \varphi$$
$$= \frac{1}{c_{1,\infty}^2}\left[\frac{\partial^2 \varphi}{\partial t^2} + 2\frac{\partial \varphi}{\partial r}\frac{\partial^2 \varphi}{\partial r \partial t} + \left(\frac{\partial \varphi}{\partial r}\right)^2\frac{\partial^2 \varphi}{\partial r^2}\right] \tag{2.11}$$

The kinematic boundary condition at the bubble wall $R(t)$ is [1].

$$u(R,t) = \left.\frac{\partial \varphi}{\partial r}\right|_{r=R} = \frac{dR}{dt} \tag{2.12}$$

The pressure condition at the bubble wall is [1].

$$p_B(t) = p_i(t) - \frac{1}{R}\left(2\sigma + 4\mu_1\frac{dR}{dt}\right) \tag{2.13}$$

with

$$p_i(t) = \left(p_\infty + \frac{2\sigma}{R_0}\right)\left(\frac{R_0}{R}\right)^{3\kappa} \tag{2.14}$$

Where σ is the surface tension coefficient, R_0 is the equilibrium radius of the bubble, κ is the polytropic index, and μ_1 is the liquid dynamic viscosity.

The enthalpy h_B on the liquid side of the bubble wall is [1].

$$h_B = \frac{c_{1,\infty}^2}{n_1 - 1}\left[\left(\frac{p_B + B}{p_\infty + B}\right)^{\frac{n_1 - 1}{n_1}} - 1\right] \tag{2.15}$$

Introducing R_0 and U as the characteristic length and velocity, respectively. The following dimensionless parameters are defined:

Dimensionless distance

$$r_* = \frac{r}{R_0} \tag{2.16}$$

Dimensionless bubble radius

$$R_* = \frac{R}{R_0} \tag{2.17}$$

Dimensionless time

$$t_* = \frac{U}{R_0} t \tag{2.18}$$

Dimensionless velocity potential

$$\varphi_* = \frac{\varphi}{R_0 U} \tag{2.19}$$

Dimensionless enthalpy

$$h_* = \frac{h}{U^2} \tag{2.20}$$

Dimensionless speed of acoustic

$$c_* = \frac{c_1}{c_{1,\infty}} \tag{2.21}$$

Mach number

$$\varepsilon = \frac{U}{c_{1,\infty}} \tag{2.22}$$

Substituting Eqs. (2.16)–(2.22) into Eqs. (2.10), (2.11), and (2.15), the dimensionless equations are [1].

$$\frac{\partial \varphi_*}{\partial t_*} + \frac{1}{2}\left(\frac{\partial \varphi_*}{\partial r_*}\right)^2 + h_* = 0 \tag{2.23}$$

$$\left[1 - \varepsilon^2 (n_1 - 1)\left(\frac{\partial \varphi_*}{\partial t_*} + \frac{1}{2}\left(\frac{\partial \varphi_*}{\partial r_*}\right)^2\right)\right]\nabla^2 \varphi_*$$
$$= \varepsilon^2\left[\frac{\partial^2 \varphi_*}{\partial t_*^2} + 2\frac{\partial \varphi_*}{\partial r_*}\frac{\partial^2 \varphi_*}{\partial r_* \partial t_*} + \left(\frac{\partial \varphi_*}{\partial r_*}\right)^2 \frac{\partial^2 \varphi_*}{\partial r_*^2}\right] \tag{2.24}$$

$$h_{\mathrm{B}*} = \frac{1}{\varepsilon^2}\frac{1}{n_1-1}\left[\left(\frac{p_\mathrm{B}+B}{p_\infty+B}\right)^{\frac{n_1-1}{n_1}}-1\right] \tag{2.25}$$

The flow field outside the bubble is divided into an inner field near the bubble and an outer field far from the bubble. According to perturbation theory, the velocity potential and enthalpy for the inner field are expressed as

$$\varphi_* = \varphi_0 + \varepsilon\varphi_1 + \varepsilon^2\varphi_2 + \cdots \tag{2.26}$$

$$h_* = h_0 + \varepsilon h_1 + \varepsilon^2 h_2 + \cdots \tag{2.27}$$

For the outer field

$$\varphi_* = \phi_0 + \varepsilon\phi_1 + \varepsilon^2\phi_2 + \cdots \tag{2.28}$$

$$h_* = H_0 + \varepsilon H_1 + \varepsilon^2 H_2 + \cdots \tag{2.29}$$

For the first-order form of the velocity potential ($\varphi_* = \varphi_0 + \varepsilon\varphi_1$) and enthalpy ($h_* = h_0 + \varepsilon h_1$), Prosperetti and Lezzi [1, 3] proposed a first-order accurate bubble dynamics equation based on the method of matched asymptotic expansions [3].

$$\begin{aligned}
&\left[1-(1+\lambda)\varepsilon\dot{R}_*\right]R_*\ddot{R}_* + \frac{3}{2}\left[1-\left(\frac{1}{3}+\lambda\right)\varepsilon\dot{R}_*\right]\dot{R}_*^2 \\
&= \left[1+(1-\lambda)\varepsilon\dot{R}_*\right]h_{\mathrm{B}*} + \varepsilon R_*\dot{h}_{\mathrm{B}*}
\end{aligned} \tag{2.30}$$

where λ is an arbitrary constant of order less than $1/\varepsilon$. $\dot{R}_*$ is the first derivative of R_* with t_*, and $\ddot{R}_*$ is the second derivative.

For the first-order form of the velocity potential ($\varphi_* = \varphi_0 + \varepsilon\varphi_1 + \varepsilon^2\varphi_2$) and enthalpy ($h_* = h_0 + \varepsilon h_1 + \varepsilon^2\varphi_2$), the second-order accurate bubble dynamics equation is [3].

$$\begin{aligned}
&\left[1-(1+\lambda)\varepsilon\dot{R}_* + \left(\frac{14}{5}+2\lambda+\theta\right)\varepsilon^2\dot{R}_*^2\right]R_*\ddot{R}_* \\
&+ \frac{3}{2}\left[1-\left(\frac{1}{3}+\lambda\right)\varepsilon\dot{R}_* + \left(\frac{16}{15}+\frac{4}{3}\lambda+\theta\right)\varepsilon^2\dot{R}_*^2\right]\dot{R}_*^2 + \varepsilon^2 R_*^2\dddot{R}_* \\
&= \left[1+(1-\lambda)\varepsilon\dot{R}_* + \theta\varepsilon^2\dot{R}_*^2\right]h_{\mathrm{B}*} + \varepsilon R_*\left[1-(1+\lambda)\varepsilon\dot{R}_*\right]\dot{h}_{\mathrm{B}*}
\end{aligned} \tag{2.31}$$

where θ is an arbitrary constant of order less than $1/\varepsilon$.

Under the influence of an external dimensionless acoustic field G_0'', the first-order and second-order accurate equations become [3].

$$\left[1-(1+\lambda)\varepsilon\dot{R}_*\right]R_*\ddot{R}_*+\frac{3}{2}\left[1-\left(\frac{1}{3}+\lambda\right)\varepsilon\dot{R}_*\right]\dot{R}_*^2$$

$$=\left[1+(1-\lambda)\varepsilon\dot{R}_*\right]\left(h_{B*}+2G_0''\right)+\varepsilon R_*\left(h_{B*}+2G_0'''\right) \tag{2.32}$$

$$\left[1-(1+\lambda)\varepsilon\dot{R}_*+\left(\frac{14}{5}+2\lambda+\theta\right)\varepsilon^2\dot{R}_*^2\right]R_*\ddot{R}_*$$

$$+\frac{3}{2}\left[1-\left(\frac{1}{3}+\lambda\right)\varepsilon\dot{R}_*+\left(\frac{16}{15}+\frac{4}{3}\lambda+\theta\right)\varepsilon^2\dot{R}_*^2\right]\dot{R}_*^2+\varepsilon^2 R_*^2\ddot{R}_*^2$$

$$=\left[1+(1-\lambda)\varepsilon\dot{R}_*+\theta\varepsilon^2\dot{R}_*^2\right]\left(h_{B*}+2G_0''\right) \tag{2.33}$$

$$+\varepsilon R_*\left[1-(1+\lambda)\varepsilon\dot{R}_*\right](h_{B*}+2G_0''')$$

$$+\varepsilon^2\left[2G_0'''R_*\ddot{R}_*+G_0^{IV}R_*^2+g_2'\right]+O\left(\varepsilon^3\right)$$

where G_0''' is the derivative of G_0'' with t_*, G_0^{IV} is the derivative of G_0''' with t_*. The dimensionless acoustic field G_0'' is defined as [3].

$$G_0''=\eta\sin\left[2\pi f t_*\right] \tag{2.34}$$

$$\eta=\frac{p_A}{U^2\rho_{1,\infty}} \tag{2.35}$$

where p_A is the amplitude of the external acoustic field, η is the dimensionless amplitude, and f is the frequency of the acoustic field.

2.1.2　Oscillation Characteristics

This section introduces the bubble oscillation characteristics under high Mach numbers using first-order and second-order equations. The amplitude and frequency of the acoustic field are 160 kPa and 30 kHz, the equilibrium radius is 3×10^{-6} m. Figure 2.1 illustrates the evolution of the dimensionless bubble radius over dimensionless time in an acoustic field. The black solid line represents the prediction from the first-order equation, while the red dashed line represents the prediction from the second-order equation. There is a noticeable difference between the two predictions. Defining a bubble period as the interval between two successive collapses to the minimum radius. After the first collapse, the second-order equation yields substantially larger local maximum radii and markedly longer periods than the first-order equation. The reason for this discrepancy is that the first-order equation

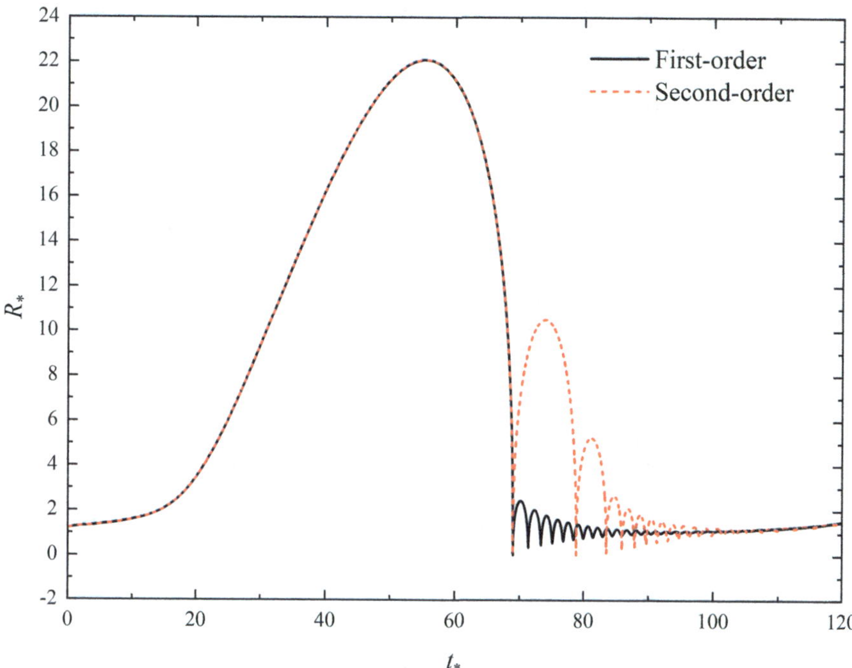

Fig. 2.1 Variation of the dimensionless bubble radius with dimensionless time

overestimates energy dissipation, resulting in small predicted local maximum bubble radii and short periods.

Figure 2.2 depicts the evolution of the dimensionless bubble wall velocity and dimensionless bubble wall acceleration over time. The amplitude and frequency of the acoustic field are 160 kPa and 30 kHz, the equilibrium radius is 3×10^{-6} m. In Fig. 2.2a, $\dot{R}_{\mathrm{reb}*}$ denotes the dimensionless velocity during the bubble rebound stage, and $\dot{R}_{\mathrm{col}*}$ denotes the dimensionless bubble wall velocity during the collapse stage. The differences between the dimensionless bubble wall velocities predicted by the first-order and second-order equations are quite pronounced, particularly in the maximum dimensionless collapse velocity and the maximum dimensionless rebound velocity. Specifically, the second-order equation predicts a maximum dimensionless collapse velocity of -1956.0 and a maximum dimensionless rebound velocity of 2006.9, whereas the first-order equation predicts values of -2388.8 and 310.4, respectively. Due to the significant energy dissipation, the differences in the predicted bubble wall velocities between the two equations diminish considerably after the first rebound. In Fig. 2.2b, the most substantial difference between the predictions of the first-order and second-order equations lies in the maximum dimensionless bubble wall acceleration ($R_{\mathrm{max}*}$). The first-order equation predicts a maximum dimensionless bubble wall acceleration of 11.7×10^7, while the second-order equation predicts 4.3×10^7.

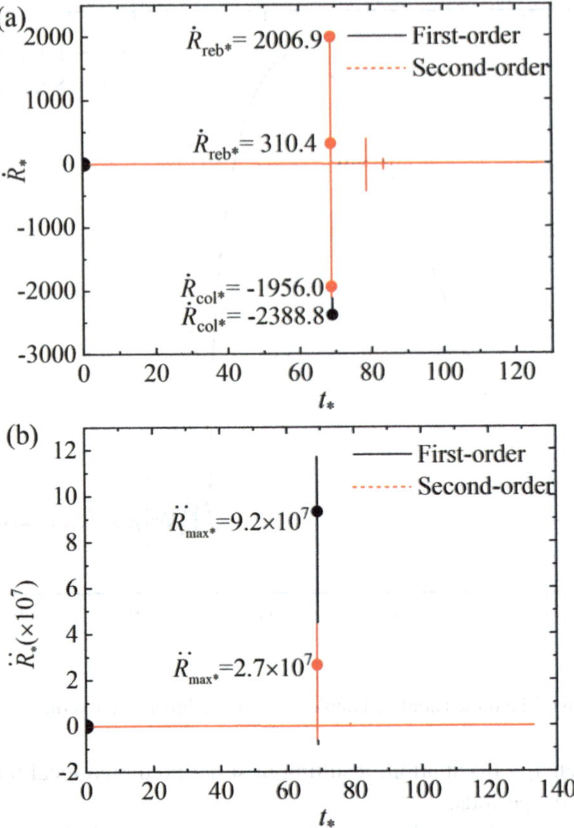

Fig. 2.2 Evolution of dimensionless bubble wall velocity and dimensionless bubble wall acceleration with time. (**a**) Dimensionless bubble wall velocity. (**b**) Dimensionless bubble wall acceleration

2.1.3 Dissipated Power

The total dissipated power (E_{tol}) of a bubble consists of radiation dissipated power (E_{r}), thermal dissipated power (E_{th}), and viscous dissipated power (E_{v}). They are expressed as [4].

$$E_{\mathrm{r}} = \frac{4\pi}{\tau c_{1,\infty}} \int_0^\tau R^2 \dot{R}\left(\dot{R}p_\infty - \frac{1}{2}\rho_1 \dot{R}^3 - \rho_1 R\dot{R}\,\dddot{R} \right)\mathrm{d}t \tag{2.36}$$

$$E_{\mathrm{th}} = f\int_0^\tau \left[-p_{\mathrm{i}}\left(1+\frac{\dot{R}}{c_{1,\infty}}\right) + \frac{R}{c_\infty}\frac{\mathrm{d}p_{\mathrm{i}}}{\mathrm{d}t} \right]\frac{\partial V}{\partial t}\,\mathrm{d}t \tag{2.37}$$

$$E_{\mathrm{v}} = f\int_0^\tau 16\pi\mu\left(\dot{R}\dot{R}^2 + \frac{R^2 \dot{R}\,\ddot{R}}{c_{1,\infty}} \right)\mathrm{d}t \tag{2.38}$$

where V is volume of the bubble. $\dot{R}$ and $\ddot{R}$ are the first and second order derivatives of R with respect to t, respectively.

Figure 2.3 illustrates the dissipation power the bubble. The equilibrium radius is 3×10^{-6} m. At large nondimensional acoustic amplitudes, the differences in both the acoustic and the thermal dissipation power predicted by the first- and second-order equations, and these differences grow as the acoustic frequency decreases. Both precision equations remain very close in the prediction of the viscous dissipation power, indicating that liquid compressibility has little effect on viscous dissipation.

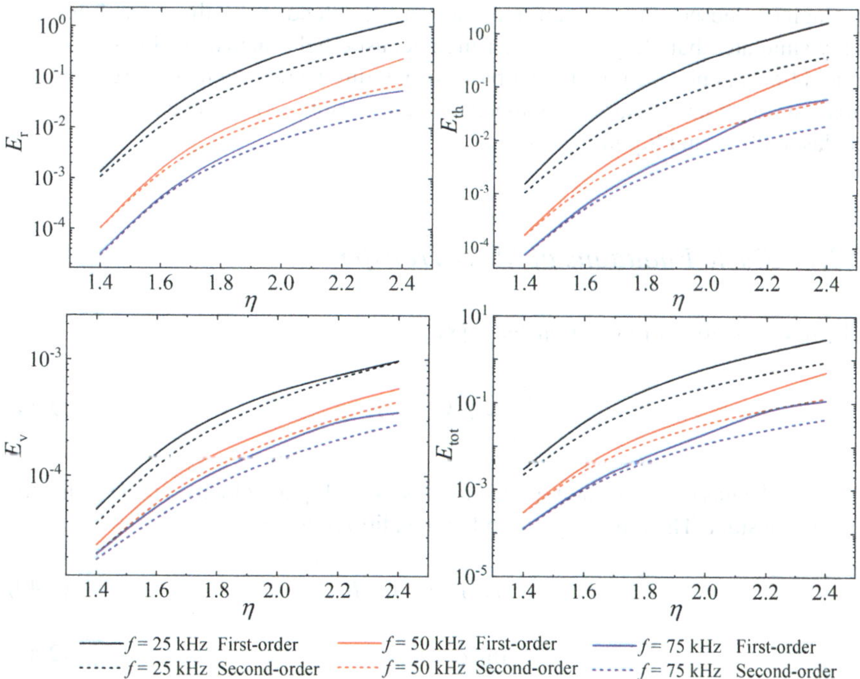

Fig. 2.3 Dissipated power of bubble. (**a**) Radiation dissipated power. (**b**) Thermal dissipated power. (**c**) viscous dissipated power. (**d**) Total dissipated power

In addition, at high nondimensional amplitudes, the total dissipation power from the first-order equations exceeds that of the second-order model, and this gap becomes more pronounced at lower acoustic frequencies.

2.2 Mass Transfer Effect of Bubble Wall

The mass transfer effect of bubbles occurs when gas inside a bubble diffuses or convects through the bubble wall due to concentration gradients and the surrounding flow field as the bubble moves in a liquid. Blake [5] provided an initial theoretical analysis of mass transfer at the bubble wall, assuming small-amplitude sinusoidal bubble motion. Later, Epstein and Plesset [6] treated the bubble wall and surrounding liquid motion as a slow process, neglecting convection, and solved the mass transfer equation independently of the bubble motion equation. Hsieh and Plesset [7] introduced a convection term into the mass transfer equation by expanding the boundary conditions as a Taylor series around the equilibrium position of the bubble wall. Eller and Flynn [8] performed a mass transfer boundary layer analysis based on the thin diffusion layer approximation, eliminated the limitations of the small sinusoidal oscillation assumption, and developed a complete threshold theory by decoupling the mass transfer equation and the bubble motion equation through time averaging. Subsequently, Crum and Hansen [9] extended the theory and numerically simulated bubble growth under high acoustic field intensities. Finally, Zhang et al. [10, 11] enhanced the model by incorporating factors such as energy dissipation, intra-bubble pressure inhomogeneity, and compressibility, extending it to viscoelastic fluids and liquid metals.

2.2.1 Basic Equations of Mass Transfer

The basic equation of mass transfer is [8].

$$\frac{\partial c}{\partial t} + \boldsymbol{u} \cdot \nabla c = D\nabla^2 c \tag{2.39}$$

here c is the gas concentration in the liquid, $\boldsymbol{u}$ is the liquid velocity, and D is the diffusion constant. The initial and boundary conditions are [8].

$$c(r,0) = c_i, \; r > R \tag{2.40}$$

$$\lim_{r \to \infty} c(r,t) = c_i \tag{2.41}$$

$$c(R,t) = c_s, \; t > 0 \tag{2.42}$$

where c_i denotes the initial uniform concentration of gas in the liquid and the concentration of gas in the liquid at infinity. c_s denotes the concentration of gas in the liquid at the bubble wall.

In this section, the equation of the bubble motion utilizes the Keller's equation [12].

$$\left(1-\frac{\dot{R}}{c_1}\right)R\ddot{R}+\frac{3}{2}\left(1-\frac{\dot{R}}{3c_1}\right)$$
$$=\left(1+\frac{\dot{R}}{c_1}\right)\frac{p_{ext}(R,t)-p_s(t)}{\rho_1}+\frac{R}{\rho_1 c_1}\frac{d\left[p_{ext}(R,t)-p_s(t)\right]}{dt} \tag{2.43}$$

with

$$p_{ext}(R,t)=p_{in}-\frac{2\sigma}{R}-\frac{4\mu_1\dot{R}}{R} \tag{2.44}$$

$$p_{in}=p_{in,eq}\left(\frac{R_0}{R}\right)^{3\kappa}-\frac{4\mu_{th}\dot{R}}{R} \tag{2.45}$$

$$p_{in,eq}=p_0+\frac{2\sigma}{R} \tag{2.46}$$

$$p_s(t)=p_0+p_A\cos(\omega t) \tag{2.47}$$

where p_{in} and $p_{in,\,eq}$ are the instantaneous pressure and the equilibrium pressure inside the bubble, respectively. σ is the surface tension. μ_{th} is the effective thermal viscosity. p_0 is the ambient pressure. p_A and ω is the amplitude and frequency of the external acoustic field, respectively.

Epstein and Plesset [6] assumed that the motion of both the bubble wall and the surrounding liquid during the diffusion of gases through the bubble wall is relatively slow, and that the interfacial area of the bubble is the main factor affecting the diffusion effect. Therefore, the convection term in the mass transfer equation is neglected, and the mass transfer equation is no longer linked to the bubble wall motion equation and can be solved independently. Eq. (2.39) after neglecting the convective term reads [6].

$$\frac{\partial c}{\partial t}=D\nabla^2 c \tag{2.48}$$

By neglecting the small change in c_s with the change in bubble radius, the solution for the change in the molar mass of gas inside the bubble can be easily solved, i.e., the final solution of Epstein's model is [6].

$$\frac{dn}{dt} = 4\pi R^2 D\left(c_{\mathrm{i}} - c_{\mathrm{s}}\right)\left\{\frac{1}{R} + \frac{1}{(\pi D t)^{\frac{1}{2}}}\right\} \tag{2.49}$$

where n is the molar mass of gas. It can be found that the threshold condition for the bubble grows is $c_{\mathrm{i}} > c_{\mathrm{s}}$. If the liquid is supersaturated with gas, the bubble grows, and conversely the bubble dissolve.

2.2.2 Mass Transfer Mode Driven by Acoustic Field

Building on Epstein's work [6], Eller and Flynn [8] considered convective effects and decoupled mass transfer from bubble wall motion by treating them as separate problems. During one oscillation period, a bubble oscillates much faster than the rate of mass transfer, resulting in negligible gas transfer across the bubble wall. Thus, the effect of mass transfer on bubble motion can be neglected, allowing the process to be treated as quasi-steady state. The Eller [8] model is described below.

The mathematical derivation begins by defining an unknown variable $\vartheta(r,t)$ and substituting it into Eq. (2.39) [8].

$$\frac{\partial c}{\partial t} + \left(\frac{\partial \vartheta}{\partial t} + \frac{R^2 \dot{R}}{r^2}\frac{\partial \vartheta}{\partial r}\right)\frac{\partial c}{\partial \vartheta} = \frac{D}{r^2}\frac{\partial}{\partial r}\left(r^2 \frac{\partial c}{\partial r}\right) \tag{2.50}$$

By setting the coefficient of $\dfrac{\partial \vartheta}{\partial t}$ to zero, we obtain

$$\frac{\partial \vartheta}{\partial t} + \frac{R^2 \dot{R}}{r^2}\frac{\partial \vartheta}{\partial r} = 0 \tag{2.51}$$

with

$$\vartheta = \frac{1}{3}\left[r^3 - R^3(t)\right] \tag{2.52}$$

Two other variables U and τ are define as [8].

$$\frac{\partial U}{\partial \vartheta} = c(\vartheta,t) - c_{\mathrm{i}} \tag{2.53}$$

$$\tau = \int_0^t R^4(t')\,dt' \tag{2.54}$$

The variables U and τ simplify the initial and boundary conditions, transforming the mass transfer equation into [8].

$$\left(1+\frac{3\vartheta}{R^3}\right)^{\frac{4}{3}}\frac{\partial^2 U}{\partial\vartheta^2}=\frac{1}{D}\frac{\partial U}{\partial\tau} \tag{2.55}$$

With the following conditions

$$U(\vartheta,0)=0 \tag{2.56}$$

$$\lim_{\vartheta\to\infty}\frac{\partial U}{\partial\vartheta}=0 \tag{2.57}$$

$$\left.\frac{\partial U}{\partial\vartheta}\right|_{\vartheta=0}=c_s-c_i \tag{2.58}$$

Since the gas mass transfer during a single oscillation is neglected, the molar mass n of gas inside the bubble remains constant. Assuming isothermal conditions, the ideal gas law yield [8].

$$p_g R^3 = p_{in,eq} R_0^3 \tag{2.59}$$

where p_g is the instantaneous pressure inside bubble. According to Henry's law and the pressure equilibrium relation Eq. (2.46) for the effect of surface tension, the concentration c_s is [8].

$$c_s = c_0\left(1+\frac{2\sigma}{R_0 p_0}\right)\left(\frac{R_0}{R}\right)^3 \tag{2.60}$$

where c_0 is the gas concentration at ambient pressure.

The boundary condition Eq. (2.58) becomes [8].

$$\left.\frac{\partial U}{\partial\vartheta}\right|_{\vartheta=0}=c_0\left(1+\frac{2\sigma}{R_0 p_0}\right)\left(\frac{R_0}{R}\right)^3-c_i=F(\tau) \tag{2.61}$$

Expanding the terms in parentheses on the left-hand side of Eq. (2.55) in a power series yields the zero-order accuracy equation [8].

$$\frac{\partial^2 U_0}{\partial\vartheta^2}-\frac{1}{D}\frac{\partial U_0}{\partial\tau}=0 \tag{2.62}$$

With conditions:

$$U_0\left(\vartheta,0\right)=0 \tag{2.63}$$

$$\lim_{\vartheta\to\infty}\frac{\partial U_0}{\partial\vartheta}=0 \tag{2.64}$$

$$\left.\frac{\partial U_0}{\partial\vartheta}\right|_{\vartheta=0}=F\left(\tau\right) \tag{2.65}$$

And the first-order accuracy equation [8].

$$\frac{\partial^2 U_1}{\partial\vartheta^2}-\frac{\partial U_1}{\partial\tau}=-\frac{4D\vartheta}{R^3}\frac{\partial^2 U_0}{\partial\vartheta^2}=-W\left(\vartheta,\tau\right) \tag{2.66}$$

with conditions

$$U_1\left(\vartheta,0\right)=0 \tag{2.67}$$

$$\lim_{\vartheta\to\infty}\frac{\partial U_1}{\partial\vartheta}=0 \tag{2.68}$$

$$\left.\frac{\partial U_1}{\partial\vartheta}\right|_{\vartheta=0}=0 \tag{2.69}$$

where $W(\vartheta,\tau)$ is a function of ϑ and τ.

After mathematical calculations such as Laplace transforms, the solutions of the zero and first order equations are obtained as [8].

$$\chi_0=-4\left(\pi D\right)^{\frac{1}{2}}\int_0^\tau \tau'^{-\frac{1}{2}}F\left(\tau-\tau'\right)\mathrm{d}\tau \tag{2.70}$$

$$\chi_1=32D\int_0^\tau\frac{\mathrm{d}\tau'\left(\tau-\tau'\right)^{\frac{1}{2}}}{R^3\left(\tau'\right)}\times\int_0^\tau F\left(\tau''\right)\frac{\mathrm{d}}{\mathrm{d}\tau''}\left[\frac{\left(\tau-\tau'\right)^{\frac{1}{2}}}{\tau-\tau''}\right]\mathrm{d}\tau'' \tag{2.71}$$

where $\chi=n-n_\mathrm{i}$ denote the amount of change in molar mass.

According to the time-averaging method, the bubble radius $R(t)$ is an assumed to be a periodic function with a period T_b. The time-averaged term for the bubble radius is expressed as

$$\langle R\rangle=\frac{1}{T_\mathrm{b}}\int_0^{T_\mathrm{b}}R^4\mathrm{d}t \tag{2.72}$$

where $\langle\bullet\rangle$ denotes the time-averaged term. The function $F(\tau)$ can be expressed as a combination of a time-averaged term F_0 and oscillatory terms, which do not affect the molar quantity variation over large time scales [8]. For the time-averaged term F_0, the equation is [8]

$$F_0 = \frac{1}{\tau_0}\int_0^{\tau_0} R^4 F(\tau)\,\mathrm{d}\tau = \frac{1}{\tau_0}\int_0^{T_b} R^4 F(t)\,\mathrm{d}t$$

$$= c_0\left[\frac{\left\langle (R/R_0)^4 (p_g/p_0)\right\rangle}{\left\langle (R/R_0)^4\right\rangle} - \frac{c_i}{c_0}\right] \tag{2.73}$$

$$\tau_0 = \int_0^{T_b} R^4\,\mathrm{d}t = R^4 T_b\left(R/R_0\right)^4 \tag{2.74}$$

where τ_0 is the time-averaged term of the time variable τ.

The rate of change of the molar mass of the gas is [8].

$$\frac{\mathrm{d}n}{\mathrm{d}t} = 4\pi R^2 J = 4\pi D R^4 \left.\frac{\partial^2 U}{\partial \vartheta^2}\right|_{\vartheta=0} \tag{2.75}$$

where J is diffusion flux. Substituting the mass transfer equation Eq. (2.51) into Eq. (2.75) at $\vartheta = 0$ yields

$$\frac{\mathrm{d}n}{\mathrm{d}t} = 4\pi \left.\frac{\partial U}{\partial t}\right|_{\vartheta=0} \tag{2.76}$$

Substituting the solutions Eqs. (2.70)–(2.71) and the time-averaged terms Eqs. (2.73)–(2.74) into Eq. (2.76)

$$\frac{\mathrm{d}n}{\mathrm{d}t} = 4\pi R_0 D c_0\left[\left\langle R/R_0 + R_0\right\rangle\left(\frac{\left\langle (R/R_0)^4\right\rangle}{\pi D t}\right)^{\frac{1}{2}}\right]\left[\frac{c_i}{c_0} - \frac{\left\langle (R/R_0)^4 (p_g/p_0)\right\rangle}{\left\langle (R/R_0)^4\right\rangle}\right] \tag{2.77}$$

When the change in gas concentration inside the bubble is zero, the threshold condition is obtained from Eq. (2.77) as

$$\frac{c_i}{c_0} = \frac{\left\langle (R/R_0)^4 (p_g/p_0)\right\rangle}{\left\langle (R/R_0)^4\right\rangle} \tag{2.78}$$

Based on Eller's model [8], Crum [9, 13] extended the analysis beyond the isothermal process limitation by incorporating a polytropic process. Zhang and Li [10] further refined Crum's threshold theory by accounting for surface tension and other factors, enhancing its accuracy. The ideal gas equation of state is given by [13].

$$P_g = P_{in,eq}\left(R/R_0\right)^{3\kappa} \tag{2.79}$$

$$P_{in,eq}V_0 = nR_gT_0 \tag{2.80}$$

where R_g is the gas constant. V_0 and T_0 are the initial volume and temperature of the bubble, respectively.

The solution to the mass transfer equation is [13].

$$\frac{dR_0}{dt} = \frac{DR_gT_0c_0}{R_0P_0}\left[R/R_0 + R_0\frac{\left(\left\langle\left(R/R_0\right)^4\right\rangle\right)^{\frac{1}{2}}}{\left\langle\left(R/R_0\right)^4\right\rangle}\right]$$

$$\times\left(1+\frac{4\sigma}{3R_0P_0}\right)^{-1}\left[\frac{c_i}{c_0} - \frac{\left\langle\left(R/R_0\right)^4\left(P_{in}/P_0\right)\right\rangle}{\left\langle\left(R/R_0\right)^4\right\rangle}\right] \tag{2.81}$$

The time-averaged terms in Eq. (2.81) require a solution to the bubble motion equation for closure. Using the perturbation method, the solution to Keller's equation Eq. (2.43) is [10].

$$R/R_0 = 1 + \alpha\left(\frac{P_A}{P_0}\right)\cos\left(\omega t + \delta\right) + \alpha^2 K\left(\frac{P_A}{P_0}\right)^2 \tag{2.82}$$

where

$$\alpha = -\frac{P_0}{\rho_1 R_0^2 M}\left[\frac{\dfrac{R_0^2\omega^2}{c_1^2}+1}{\left(\omega_0^2-\omega^2\right)^2+4\beta_{tol}^2\omega^2}\right]^{\frac{1}{2}} \tag{2.83}$$

$$\delta = \tan^{-1}\left[\frac{\dfrac{R_0\omega}{c_1}\left(\omega_0^2-\omega^2\right)-2\beta_{tol}\omega}{\omega_0^2-\omega^2+\dfrac{2R_0\beta_{tol}}{c_1}\omega^2}\right] \tag{2.84}$$

$$K = \frac{\left(3\kappa + 1 - \dfrac{\rho\omega^2 R_0^2}{3\kappa p_0}\right)/4 + \left(\dfrac{\sigma}{2R_0 p_0}\right)(3\kappa + 1 - 2/3\kappa)}{1 + \left(\dfrac{2\sigma}{R_0 p_0}\right)(1 - 1/3\kappa)} \tag{2.85}$$

with

$$M = 1 + \frac{4\left(\mu_1 + \mu_{\mathrm{th}}\right)}{\rho_1 R_0^2}\frac{R_0}{c_1} \tag{2.86}$$

$$\omega_0^2 = \left[3\kappa\left(p_0 + \frac{2\sigma}{R_0}\right) - \frac{2\sigma}{R_0}\right]\frac{1}{\rho_1 R_0^2 M} \tag{2.87}$$

$$\beta_{\mathrm{tol}} = \frac{2\left(\mu_1 + \mu_{\mathrm{th}}\right)}{\rho_1 R_0^2 M} + \frac{R_0}{2c_1}\omega_0^2 \tag{2.88}$$

The natural frequency ω_0 and damping constant β_{tol} are given directly here, and they will be discussed in Sect. 2.3. The time-averaged term in Eq. (2.81) is solved by Eqs. (2.82)–(2.88). The obtained time-averaged terms $\langle R/R_0 \rangle$, $\langle (R/R_0)^4 \rangle$ and $\langle (R/R_0)^4(p_g/p_0) \rangle$ are [14].

$$\langle R/R_0 \rangle = 1 + K\alpha^2\left(\frac{p_A}{p_0}\right)^2 \tag{2.89}$$

$$\langle (R/R_0)^4 \rangle = 1 + \alpha^2(3 + 4K)\left(\frac{p_A}{p_0}\right)^2 \tag{2.90}$$

$$\langle (R/R_0)^4(p_g/p_0) \rangle =$$
$$\left(1 + \frac{2\sigma}{R_0 p_0}\right)\left[1 + \frac{3\alpha^2(\kappa-1)(3\kappa-4)}{4}\left(\frac{p_A}{p_0}\right)^2 + K\alpha^2(4-3\kappa)\left(\frac{p_A}{p_0}\right)^2\right] \tag{2.91}$$

At $dR/dt = 0$, the threshold value of the acoustic pressure amplitude is [14].

$$p_T^2 = \frac{p_0^2}{\alpha^2}\frac{1 + \dfrac{2\sigma}{R_0 p_0} - \dfrac{c_i}{c_0}}{(4K+3)\dfrac{c_i}{c_0} - \left[\dfrac{3}{4}(\kappa-1)(3\kappa-4) + (4-3\kappa)K\right]\left(1 + \dfrac{2\sigma}{p_0 R_0}\right)} \tag{2.92}$$

2.2.3 Mass Transfer Under Acoustic Field of Dual Frequency

The bubble dynamics equation is based on the Keller' equation [Eq. (2.43)], the external acoustic pressure term in Eq. (2.47) is replaced by the dual-frequency counterpart [14].

$$p_s = p_0 + p_{A1}\cos(\omega_1 t) + p_{A2}\cos(\omega_2 t) \tag{2.93}$$

where p_{A1} and p_{A2} are the acoustic pressure amplitudes at frequencies ω_1 and ω_2, respectively, and it is assumed the $\omega_1 < \omega_2$.

Using a perturbation expansion, the solution of the bubble radius under dual-frequency excitation is [14].

$$\frac{R}{R_0} = 1 + B_2\left(\frac{p_{A1}}{p_0}\right)^2 + \left[A_{11}\cos(\omega_1 t + \delta_{11}) + A_{12}\cos(\omega_2 t + \delta_{12})\right]\frac{p_{A1}}{p_0} \tag{2.94}$$

with the coefficients defined as

$$A_{11} = -\frac{p_0}{M\rho_1 R_0^2}\left[\frac{1 + \left(\dfrac{\omega_1 R_0}{c_1}\right)^2}{\left(\omega_0^2 - \omega_1^2\right)^2 + 4\beta_{tol}^2\omega_1^2}\right]^{1/2} \tag{2.95}$$

$$\delta_{11} = \tan^{-1}\left[\frac{\dfrac{R_0\omega_1}{c_1}\left(\omega_0^2 - \omega_1^2\right) - 2\beta_{tol}\omega_1}{\omega_0^2 - \omega_1^2 + \dfrac{2R_0\beta_{tol}}{c_1}\omega_1^2}\right] \tag{2.96}$$

$$A_{12} = -\frac{p_0 p_{A2}}{M\rho_1 R_0^2 p_{A1}}\left[\frac{1 + \left(\dfrac{\omega_2 R_0}{c_1}\right)^2}{\left(\omega_0^2 - \omega_2^2\right)^2 + 4\beta_{tol}^2\omega_2^2}\right]^{1/2} \tag{2.97}$$

$$\delta_{12} = \tan^{-1}\left[\frac{\dfrac{R_0\omega_2}{c_1}\left(\omega_0^2 - \omega_2^2\right) - 2\beta_{tol}\omega_2}{\omega_0^2 - \omega_2^2 + \dfrac{2R_0\beta_{tol}}{c_1}\omega_2^2}\right] \tag{2.98}$$

$$B_2 = -\frac{\left(A_{11}^2\omega_1^2 + A_{12}^2\omega_2^2\right)}{4\omega_0^2 M} \frac{A_{11}^2 + A_{12}^2}{4\rho_1 R_0^2 \omega_0^2 M}\left[3\kappa(3\kappa+1)\left(p_0 + \frac{2\sigma}{R_0}\right) - \frac{4\sigma}{R_0}\right] \tag{2.99}$$

$$\beta_{\text{tol}} = \frac{2\mu_1}{\rho_1 R_0^2 M} + \frac{R_0}{2c_1}\omega_0^2 \tag{2.100}$$

$$M = 1 + \frac{4\mu_1}{\rho_1 R_0^2}\frac{R_0}{c_1} \tag{2.101}$$

Substituting Eqs. (2.94)–(2.101) into the time-averaged terms of Eq. (2.81) yields

$$\langle R / R_0 \rangle = 1 + B_2\left(\frac{p_{A1}}{p_0}\right)^2 \tag{2.102}$$

$$\left\langle (R / R_0)^4 \right\rangle = 1 + \left[4B_2 + 3\left(A_{11}^2 + A_{12}^2\right)\right]\left(\frac{p_{A1}}{p_0}\right)^2 \tag{2.103}$$

$$\left\langle (R / R_0)^4 \left(p_g / p_0\right)\right\rangle =$$
$$\left(1 + \frac{2\sigma}{R_0 p_0}\right)\left\{1 + \left[(4-3\kappa)B_2 + \frac{(4-3\kappa)(3-3\kappa)\left(A_{11}^2 + A_{12}^2\right)}{4}\right]\left(\frac{p_{A1}}{p_0}\right)^2\right\} \tag{2.104}$$

For simplicity, the two frequency acoustic fields are assumed to have equal amplitudes, i.e., $p_{A1} = p_{A2} = \tilde{p}$. The acoustic pressure amplitude threshold of the mass transfer in the dual-frequency acoustic field excitation is [14].

$$\tilde{p}_T^2 = \frac{\left(1 + \dfrac{2\sigma}{R_0 p_0} - \dfrac{c_i}{c_0}\right)}{\dfrac{A_{11}^2 + A_{12}^2}{p_0^2}\left[\dfrac{3c_i}{c_0} - \dfrac{3(4-3\kappa)(1-\kappa)}{4}\left(1 + \dfrac{2\sigma}{R_0 p_0}\right)\right] + \dfrac{B_2}{p_0^2}\left[\dfrac{4c_i}{c_0} - \left(1 + \dfrac{2\sigma}{R_0 p_0}\right)(4-3\kappa)\right]} \tag{2.105}$$

Figure 2.4 illustrates a comparison of the predicted thresholds of corrected mass diffusion of acoustic pressure amplitude for single- and dual-frequency acoustic excitation. In the Fig. 2.4, $\omega_1 = 5 \times 10^5$ s^{-1}, $\omega_2 = 1.5 \times 10^6$ s^{-1}, $c_i/c_0 = 1$. If $\tilde{p} > \tilde{p}_T$, the bubble grows. if $\tilde{p} < \tilde{p}_T$, the bubble dissolves. The resonant bubble radii for the two frequencies, R_{r1} and R_{r2}, are also indicated in Fig. 2.4. For single-frequency

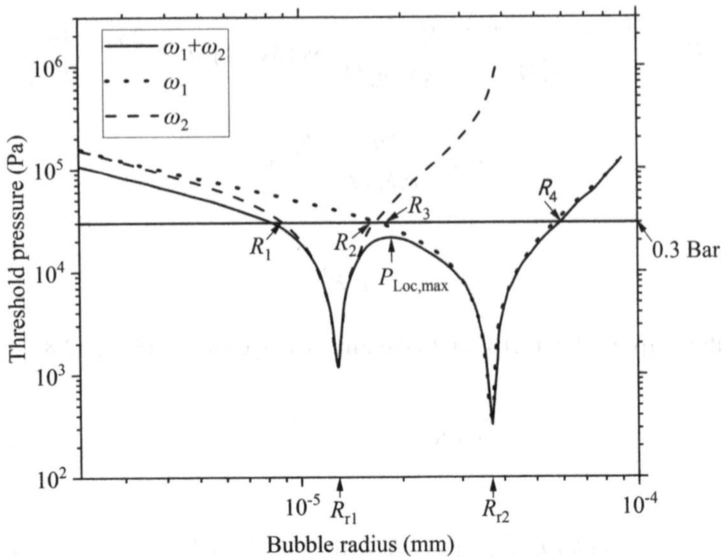

Fig. 2.4 Comparison of the predicted thresholds of rectified mass diffusion of acoustic pressure amplitude for single- and dual-frequency acoustic excitations. (Reprinted with the permission from Ref. [14] Copyright (2012) (ELSEVIER))

excitation at ω_2, the threshold curve intersects the $\tilde{p}$ at radii R_1 and R_2 ($R_1 < R_2$), while at ω_1 it intersects the $\tilde{p}$ at R_3 and R_4 ($R_3 < R_4$). For dual-frequency excitation, the differences in the threshold curves are negligible, so we approximate its intersections with the $\tilde{p}$ as still R_1 and R_4. Compared to single-frequency excitation, the dual-frequency threshold is greatly reduced within the resonant range $[R_{r1}, R_{r2}]$, and a local maximum threshold pressure $p_{LOC, Max}$ appears. If the dual-frequency amplitude exceeds $p_{Loc, Max}$ (e.g., $\tilde{p} = 0.3$ bar in Fig. 2.4), the four intersection points R_1-R_4 divide bubble behavior into five regions. Consequently, the radius range for bubble growth under dual-frequency excitation $[R_1, R_4]$ is much broader than under ω_2 single-frequency $[R_1, R_2]$ or ω_1 single-frequency $[R_3, R_4]$ excitation. Since the dual-frequency threshold in $[R_{r1}, R_{r2}]$ is low, Eq. (2.81) predicts higher bubble growth rates in this interval. When $\tilde{p} < p_{Loc,max}$, bubble responses under single- and dual-frequency excitations do not differ significantly.

2.3 Energy Dissipation Mechanism of Bubbles

2.3.1 Natural Frequency and Damping Mechanism

This section introduces the natural frequency and damping constant based on the Keller's equation. In an infinite liquid, the bubble oscillates with small amplitude. In the Keller's equation [Eq. (2.43)], the applied acoustic field is given by [15].

$$p_s(t) = p_0\left(1 + \varrho\, e^{i\omega t}\right) \tag{2.106}$$

here ϱ is the dimensionless amplitude of the acoustic field. The instantaneous bubble radius is assumed to be [15].

$$R = R_0\left(1 + x\right) \tag{2.107}$$

where x represents the dimensionless perturbation of the radius, on the same order as ϱ. Substituting Eq. (2.107), Eq. (2.43) is linearized. For small oscillations, neglecting the second-order or higher-order terms of ϱ, the inhomogeneous equation for a harmonic oscillator is obtained as by Zhang [15].

$$\ddot{x} + 2\beta_{\text{tol}}\dot{x} + \omega_0^2 x = -\alpha_0 \varrho\left(1 + \frac{i\omega R_0}{c_1}\right)\frac{e^{i\omega t}}{M} \tag{2.108}$$

Here, the total damping constant β_{tol}, natural frequency ω_0 and constant M are defined as [15].

$$\alpha_0 = \frac{p_0}{\rho_1 R_0^2} \tag{2.109}$$

$$\beta_{\text{tol}} = \beta_{\text{vis}} + \beta_{\text{th}} + \beta_{\text{ac}} = \frac{2\left(\mu_1 + \mu_{\text{th}}\right)}{\rho_1 R_0^2 M} + \frac{R_0}{2c_1}\omega_0^2 \tag{2.110}$$

$$\omega_0^2 = \left(\frac{3\kappa p_{\text{in,eq}}}{\rho_1 R_0^2} - \frac{2\sigma}{\rho_1 R_0^3}\right)\frac{1}{M} \tag{2.111}$$

$$M = 1 + \frac{R_0}{c_1}\frac{4\left(\mu_1 + \mu_{\text{th}}\right)}{\rho_1 R_0^2} \tag{2.112}$$

For the natural frequency, the contributions of liquid viscosity μ_1, thermal viscosity μ_{th}, and liquid compressibility through M are very small [15]. The total damping β_{tol} comprises viscous damping β_{vis}, thermal damping β_{th}, and acoustic damping β_{ac}, with their expressions being [15]

$$\beta_{\text{vis}} = \frac{2\mu_1}{\rho_1 R_0^2 M} \tag{2.113}$$

$$\beta_{\text{th}} = \frac{2\mu_{\text{th}}}{\rho_1 R_0^2 M} \tag{2.114}$$

$$\beta_{ac} = \frac{R_0 \omega_0^2}{2c_1} \tag{2.115}$$

For an incompressible liquid ($c_1 \rightarrow \infty$), terms such as $\dot{R}/c_1$ and R_0/c_1 vanish. Consequently, the Keller's equation [Eq. (2.43)] simplifies to the Rayleigh-Plesset equation [18], and Eq. 2.108 becomes [16].

$$\ddot{x} + \frac{4(\mu_1 + \mu_{th})}{\rho_1 R_0^2}\dot{x} + \omega_0^2 x = -\alpha_0 \varrho\, e^{i\omega t} \tag{2.116}$$

Keller and Kolodner [17] further simplified this by neglecting liquid viscosity ($\mu_1 = 0$ and $\mu_{th} = 0$), the acoustic field ($\varepsilon = 0$) and surface tension ($\sigma = 0$), reducing the equation to [15].

$$\ddot{x} + \frac{3\kappa p_0}{\rho_1 R_0^2 c_1}\dot{x} + \frac{3\kappa p_0}{\rho_1 R_0^2} x = 0 \tag{2.117}$$

This simplified equation (Eq. 2.117) accurately predicts bubble radius behavior, aligning with experimental data [17]. Keller and Miksis [12] noted that including viscosity has negligible effects. If $\mu_{th} = 0$ and $\varrho = 0$, the inhomogeneous equations for the resonator of an oscillating bubble derived by Shima [19] based on Gilmore's equation can be obtained [15].

$$\left(1 + \frac{4R_0}{\rho_1 R_0^2 c_1}\right)\mu_1\,\ddot{x} + \left(\frac{4\mu_1}{\rho_1 R_0^2 M} + \frac{R_0}{c_1}\omega_0^2\right)\dot{x} + \omega_0^2 x = 0 \tag{2.118}$$

When linearizing the Gilmore equation, the speed of acoustic is treated as constant, and second-order terms are neglected, resulting in a damping expression identical to that from the Keller's equation.

2.3.2 Thermal Effects

To fully define the model, thermodynamic parameters κ and μ_{th} in equations Eqs. (2.111) and (2.112) need to be determined. Assuming a small deviation of the quantities in the bubble from the equilibrium value [16].

$$\rho = \rho_g\left(1 + \rho_{g*}\right) \tag{2.119}$$

$$P = p_0\left(1 + \frac{2\sigma}{R_0 p_0} + p_{g*}\right) \tag{2.120}$$

$$T_g = T_\infty \left(1 + T_{g*}\right) \tag{2.121}$$

$$T_l = T_\infty \left(1 + T_{l*}\right) \tag{2.122}$$

where ρ, P and T_g are the local values of density, pressure and temperature in the gas, respectively. T_∞ and T_l are the ambient temperature and the Liquid local temperature, respectively. ρ_{g*}, p_{g*} and T_{g*} are the non-dimensional deviations from the equilibrium values of density, pressure and temperature in the gas respectively. T_{l*} is the non-dimensional deviation from the equilibrium value of the liquid temperature. The equation of state for gas is

$$P = \rho T_g \frac{R_g}{M_g} \tag{2.123}$$

where R_g and M_g are the universal gas constant and molecular weight of gas in the bubble, respectively. Based on the above relationships, the following relationships can be obtained [16].

$$\eta_{p,g} = \left(1 + \frac{2\sigma}{R_0 p_0}\right)\left(\rho_{g*} + T_{g*}\right) \tag{2.124}$$

The equations of conservation of mass, momentum and energy for the gas are [16].

$$\frac{\partial \rho_{g*}}{\partial t} + \frac{1}{r^2}\frac{\partial\left(r^2 u_g\right)}{\partial r} = 0 \tag{2.125}$$

$$\frac{\partial u_g}{\partial t} + \frac{p_0}{\rho_g}\frac{\partial p}{\partial r} = 0 \tag{2.126}$$

$$\frac{1}{r^2}\frac{\partial}{\partial r}\left(r^2 \frac{\partial T_{g*}}{\partial r}\right) + \frac{p_0 + 2\sigma / R_0}{k_g T_\infty}\frac{\partial \rho_{g*}}{\partial t} = \frac{1}{D_{g,v}}\frac{\partial T_{g*}}{\partial t} \tag{2.127}$$

The energy conservation equation for the liquid is [16].

$$\frac{1}{r^2}\frac{\partial}{\partial r}\left(r^2 \frac{\partial T_{l*}}{\partial r}\right) = \frac{1}{D_l}\frac{\partial T_{l*}}{\partial t} \tag{2.128}$$

where u_g is the gas velocity. k_g and k_l are the thermal conductivities of gas and liquid, respectively. $D_{g,v}$ and D_l are the thermal diffusivities of gas and liquid, respectively.

At the gas–liquid interface ($r = R_0$), the boundary conditions are given as follows [16].

$$T_{1*}\big|_{(r=R_0,t)} = T_{g*}\big|_{(r=R_0,t)} \tag{2.129}$$

$$-k_1 \frac{\partial T_{1*}}{\partial t}\bigg|_{(r=R_0,t)} = -k_g \frac{\partial T_{1*}}{\partial t}\bigg|_{(r=R_0,t)} \tag{2.130}$$

$$u_g\big|_{(r=R_0,t)} = dR / dt \tag{2.131}$$

Based on the above equations and boundary conditions, the expressions for κ and μ_{th} are [16, 20].

$$\mu_{th} = \frac{1}{4}\omega\rho_g R_0^2 \, Im\psi \tag{2.132}$$

$$\kappa = \frac{1}{3}\frac{\omega^2 \rho_g R_0^2}{P_{in,eq}} Re\psi \tag{2.133}$$

with

$$\psi = \frac{kf\left(\Gamma_2 - \Gamma_1\right) + \lambda_2\Gamma_2 - \lambda_1\Gamma_1}{kf\left(\lambda_2\Gamma_1 - \lambda_1\Gamma_2\right) - \lambda_1\lambda_2\left(\Gamma_2 - \Gamma_1\right)} \tag{2.134}$$

where

$$\beta_{1,2} = \left(\frac{1}{2}\gamma G_2\left\{i - G_1 \pm\left[\left(i - G_1\right)^2 + \frac{4iG_1}{\gamma}\right]\right\}\right)^{\frac{1}{2}} \tag{2.135}$$

$$\Gamma_{1,2} = i + G_1 \pm\left[\left(i - G_1\right)^2 + \frac{4iG_1}{\gamma}\right]^{\frac{1}{2}} \tag{2.136}$$

$$\lambda_i = \beta_i \coth\beta_i - 1, \; i = 1,2 \tag{2.137}$$

$$f = 1 + (1+i)\left(\frac{G_3}{2}\right)^{\frac{1}{2}} \tag{2.138}$$

$$k = \frac{k_1}{k_g} \tag{2.139}$$

$$G_1 = \frac{M_g D_{g,v} \omega}{\gamma R_g T_\infty} \tag{2.140}$$

$$G_2 = \frac{\omega R_0^2}{D_{g,v}} \tag{2.141}$$

$$G_3 = \frac{\omega R_0^2}{D_1} \tag{2.142}$$

where γ is the ratio of specific heats of the gas. G_1 reflects the ratio between mean free path and the wavelength in the gas. G_2 reflects the ratio between bubble radius and thermal penetration depth. k is the ratio of the liquid and the gas thermal conductivities. Faced with the fact that Eqs. (2.134)–(2.142) are very complex, Prosperetti [16] simplified Eqs. (2.134)–(2.142) by assuming that the angular frequency of the acoustic field is limited ($G_1 \ll 1$) and found that kf is a large value in most cases. The simplified results are given as follows [16].

$$\psi = \Theta \left[1 + (kf)^{-1} E_1 + O(kf)^{-2} \right] \tag{2.143}$$

with

$$\Theta = \frac{\Gamma_2 - \Gamma_1}{\lambda_2 \Gamma_2 - \lambda_1 \Gamma_1} \tag{2.144}$$

$$E_1 = \frac{\Gamma_1 \Gamma_2}{\Gamma_1 - \Gamma_2} \frac{(\lambda_1 - \lambda_2)^2}{\lambda_1 \Gamma_2 - \lambda_2 \Gamma_1} \tag{2.145}$$

$$\beta_1 = (1+i) \left(\frac{\gamma G_2}{2} \right)^{\frac{1}{2}} \left\{ 1 + \frac{iG_1 (\gamma - 1)}{2\gamma} + O(G_1^2) \right\} \tag{2.146}$$

$$\beta_2 = (G_1 G_2)^{\frac{1}{2}} \left[i + \frac{G_1 (\gamma - 1)}{2\gamma} + O(G_1^2) \right] \tag{2.147}$$

$$\Gamma_1 = 2 \left(i + \frac{G_1}{\gamma} \right) + O(G_1^2) \tag{2.148}$$

$$\Gamma_2 = 2G_1 \left(\frac{\gamma - 1}{\gamma} \right) + O(G_1^2) \tag{2.149}$$

If all $(kf)^{-1}$ terms with the first order and above are neglected, Eq. (2.143) reduces to

$$\psi = \Theta \tag{2.150}$$

Zhang and Li [21] investigated the validity of Prosperetti's assumption [17] that $G_1 \ll 1$, finding it inadequate in the high-frequency region (above megahertz). They attributed this failure to Prosperetti's neglect of higher-order terms in G_1. To address this limitation and better account for high-frequency effects, Zhang [20, 21] extends G_1 to third order while omitting insignificant terms in Γ_2, obtaining [20].

$$\Gamma_1 = \frac{2G_1}{\gamma} + \frac{2}{\gamma}\left(\frac{2}{\gamma}-1\right)\left(1-\frac{1}{\gamma}\right)G_1^3 + i\left[2+\frac{2}{\gamma}\left(\frac{1}{\gamma}-1\right)G_1^2\right] + O\left(G_1^4\right) \tag{2.151}$$

$$\Gamma_2 = 2(\gamma-1)G_1\left(1+\frac{iG_1}{\gamma}\right)/\gamma + O\left(G_1^3\right) \tag{2.152}$$

$$\beta_1 = \left(\frac{2\gamma G_2}{2}\right)^{\frac{1}{2}}\left\{2\left(\frac{1}{\gamma}-1\right)G_1 + \frac{2}{\gamma}\left(\frac{2}{\gamma}-1\right)\left(1-\frac{1}{\gamma}\right)G_1^3 \right.$$
$$\left. +i\left[2+\frac{2}{\gamma}\left(\frac{1}{\gamma}-1\right)G_1^2\right] + O\left(G_1^4\right)\right\}^{\frac{1}{2}} \tag{2.153}$$

$$\beta_2 = (G_1 G_2)^{\frac{1}{2}}\left\{-1-\left(\frac{2}{\gamma}-1\right)\left(1-\frac{1}{\gamma}\right)G_1^2 + \left(1-\frac{1}{\gamma}\right)G_1 i + O\left(G_1^3\right)\right\}^{\frac{1}{2}} \tag{2.154}$$

For $G_1 G_2 \ll 1$, Devin [22] assumes that the pressure inside the bubble is uniform, with small oscillations in pressure, bubble volume and temperature, and constant temperature of the liquid near the bubble, giving an expression for the thermal effect [22].

$$\gamma G_1 G_2 \psi \approx \Psi = \frac{3\gamma}{1-3(\gamma-1)iX\left[\left(\dfrac{i}{X}\right)^{\frac{1}{2}}\coth\left(\dfrac{i}{X}\right)^{\frac{1}{2}}-1\right]} \tag{2.155}$$

with

$$X = \frac{D_{g,p}}{\omega R_0^2} \tag{2.156}$$

where $D_{g,p} = D_{g,}\sqrt{\gamma}$ is the thermal diffusivity of the gas at constant pressure. Eqs. (2.132) and (2.133) can be expressed as [22].

$$\mu_{th} = \frac{p_{in,eq}}{4\omega} Im\Psi \qquad (2.157)$$

$$\kappa = \frac{1}{3} Re\,\Psi \qquad (2.158)$$

The non-dimensional thermal damping constant (δ_{th}) is defined by Devin [22] as

$$\delta_{th} = \frac{4\mu_{th}\omega}{\rho_1 R_0^2 \omega_0^2} = 2\beta_{th}\frac{\omega}{\omega_0^2} \qquad (2.159)$$

Devin [22] gave the expression of δ_{th} as follows

$$\delta_{th} = \frac{Im\Psi}{Re\,\Psi} \qquad (2.160)$$

When considering surface tension, the δ_{th} is [20].

$$\delta_{th} = \frac{Im\Psi}{Re\,\Psi - 2\sigma/R_0 p_{in,eq}} \qquad (2.161)$$

Figure 2.5 compares the thermodynamic parameters κ and β_{th} predicted by the four sets of expressions. For the multiparty index κ, all four sets of expressions yield similar predictions when $G_2 < 10$. As G_2 increases, Devin's predictions [22] increasingly diverge from the exact solution. In contrast, the predictions by Zhang [20] and those based on $G_1 \ll 1$ [16] remain closely aligned with the exact solution. For small G_2 values, predictions using $G_1 \ll 1$ [16] deviate noticeably from the exact solution. At large G_2 values, Devin's predictions show significant deviation. Zhang's results [20] demonstrate strong agreement with the exact solution across a wide range of large G_2 values.

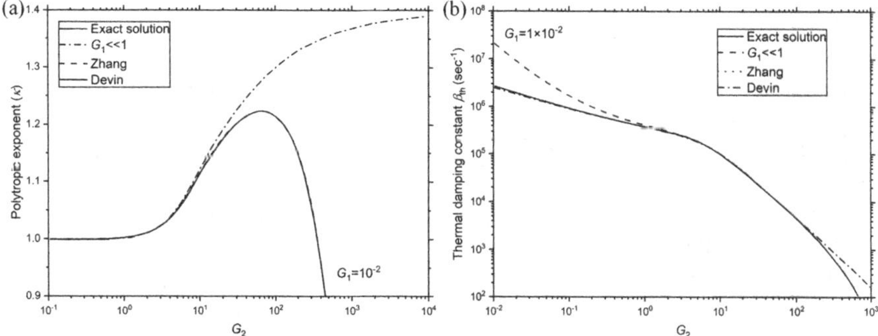

Fig. 2.5 Comparison of the thermodynamic parameters κ and β_{th} predicted by the four sets of expressions. (Reprinted with the permission from Ref. [20] Copyright (2013) (ELSEVIER))

2.3.3 Acoustic Damping

In Sect. 2.3.1, Zhang [15] provided the expression for acoustic damping in compressible fluids $\beta_{ac} = R_0\omega_0^2 / 2c_1$, which differs from Prosperetti's result [23] of $R_0\omega^2/2c_1$. Zhang and Li [24] reviewed the derivation process of Prosperetti's result [23]. Following Prosperetti [23], by dividing both sides of Eq. (2.108) by $(1 + R_0\omega/c_1)$, ignoring terms of order c^{-2}, and using the relations $x = i\omega\dot{x}$ and $\dot{x} = i\omega x$, Eq. (2.108) becomes

$$\ddot{x} + 2\left[\frac{2(\mu_1 + \mu_{th})}{\rho_1 R_0^2} + \frac{R_0\omega^2}{2c_1}\right]\dot{x} + \omega_0^2 x = -\alpha_0 \varrho\, e^{i\omega t} \tag{2.162}$$

Figure 2.6 compares the acoustic damping and total damping from Prosperetti [16] and Zhang [24]. The results from Prosperetti and Zhang differ significantly. For acoustic damping, as the equilibrium bubble radius increases, Zhang's prediction decreases, while Prosperetti's prediction increases. For small equilibrium bubble radii, the total damping predictions from both models are similar. However, for

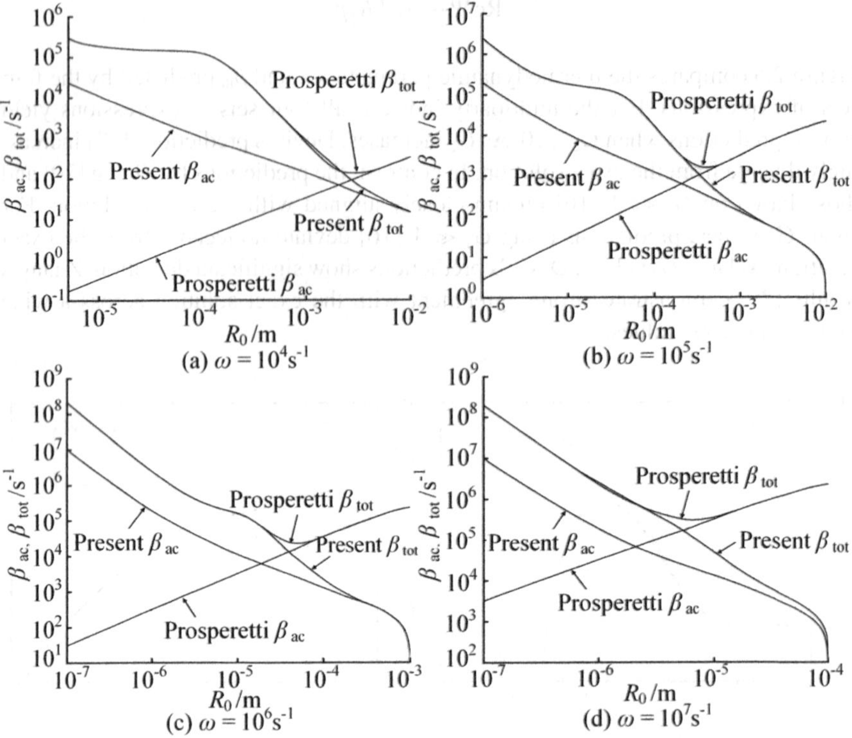

Fig. 2.6 Comparison of acoustic and total damping from Prosperetti [16] and Zhang [24] predictions. Reprinted with the permission from Ref. [24] Copyright (2012) (ELSEVIER)

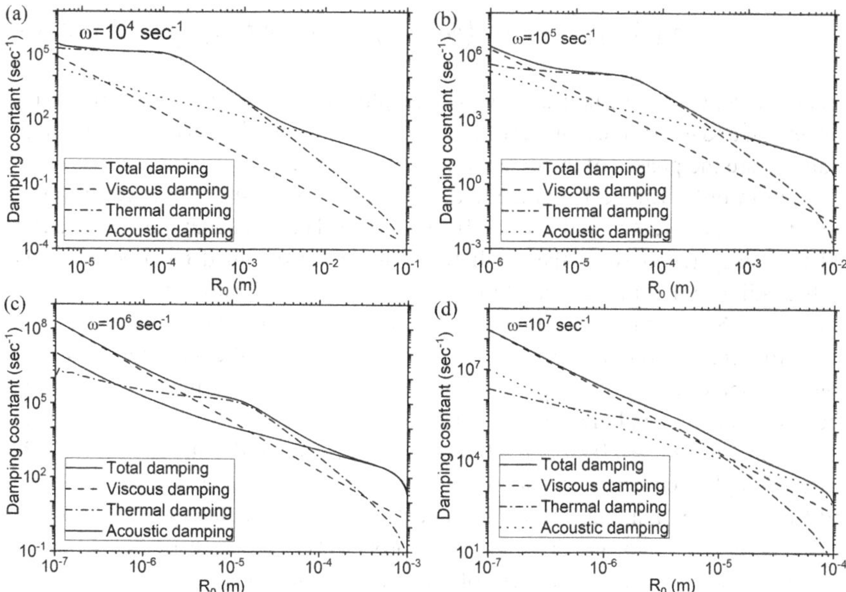

Fig. 2.7 Energy dissipation of an oscillating gas bubble in water. (Reprinted with the permission from Ref. [20] Copyright (2013) (ELSEVIER))

large radii, Prosperetti's predicted acoustic damping increases, leading to an increase in total damping. Comparing Fig. 2.6a, d, in high-frequency acoustic fields, the equilibrium bubble radius for Prosperetti's model is significantly reduced. Therefore, in high-frequency acoustic fields, Zhang's model can reduce prediction errors.

Figure 2.7 illustrate the energy dissipation of an oscillating gas bubble in water. As the equilibrium radius of the bubble increases, all damping decreases. For small bubbles, viscous damping is dominant. For large bubbles, acoustic damping is dominant. While thermal damping dominates in the intermediate region between the two regions above. As the frequency increases, the contribution of thermal damping to the total damping decreases.

2.4 Bubble Dynamics Under Dual-Frequency Acoustic Excitation

2.4.1 Bubble Response

In this section, the bubble dynamics equation is based on the Keller' equation [Eq. (2.43)], The external acoustic field pressures for the single and dual frequencies are

$$p_s(t) = p_0 \left[1 + \varepsilon_s \cos\left(\omega_s t + \varphi_s \right) \right] \tag{2.163}$$

$$p_s(t) = p_0 \left[1 + \varepsilon_1 \cos(\omega_1 t + \varphi_1) + \varepsilon_2 \cos(\omega_2 t + \varphi_2) \right] \qquad (2.164)$$

where ε_s, ε_1 and ε_2 are the non-dimensional amplitude of the external acoustic excitation. ω_s, ω_1 and ω_2 are the angular frequencies of the external acoustic excitation. φ_s, φ_1 and φ_2 are the phases of the external acoustic waves.

A fourth-order Runge-Kutta method [25] with a fixed time step is employed to solve Eqs. (2.43)–(2.46) and Eqs. (2.163)–(2.164). The frequency power spectrum and frequency response of the oscillating bubble are obtained by processing the bubble radius in the time domain through the Fast Fourier Transform.

Figure 2.8 illustrates the time-domain variation of the dimensionless bubble radius for single and dual frequencies. Figures 2.8a, b are the variation of dimensionless bubble radius for single- and dual-frequency acoustic fields, respectively. The dimensionless bubble radius is defined as $X = (R - R_0)/R_0$. In the figures, $R_0 = 10$ μm, $\omega_s = \omega_1 = 0.03\omega_0$, $\omega_2 = 0.057\omega_0$, $\varepsilon_s = \varepsilon_1 = \varepsilon_2 = 0.05$, and $p_0 = 101325$ Pa. Under a single-frequency acoustic field, the bubble oscillates at a fixed amplitude and frequency. With a dual-frequency acoustic field, the oscillatory behavior of the bubble is very complex and depends on the properties of the acoustic fields, such as the amplitude, frequency, and phase of the two acoustic fields.

Figure 2.9 demonstrates the power spectrum of bubble oscillations under the external acoustic field. Figure 2.9a, b are power of the bubble oscillations for single- and dual-frequency acoustic fields, respectively. With a single-frequency acoustic field, the power spectrum contains a main band equal to the frequency of the acoustic field as well as the first and the second harmonics. With a dual-frequency acoustic field, the power spectrum of bubble oscillations contains multiple bands. The frequencies corresponding to these bands can be expressed as a linear combination of the frequencies of the two component acoustic waves, i.e. $\omega_p = a\omega_1 + b\omega_2$ (a and b are two integers). The power spectrum of bubble oscillations exhibits three frequency bands, namely, the main resonances band, the harmonic band, and the combination resonance band. The main resonance bands are labeled $(1, 0)$ and $(0, 1)$, the

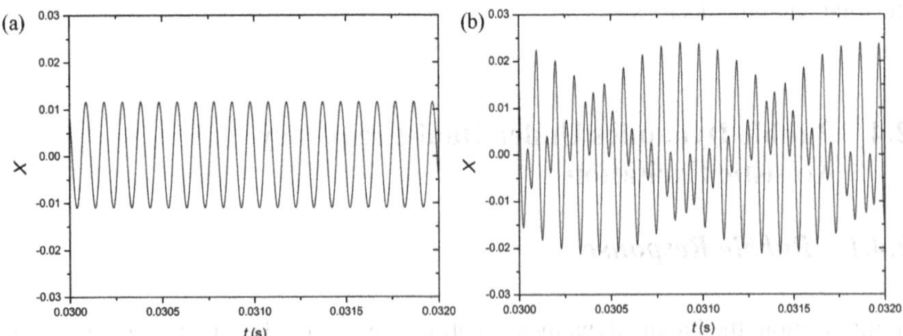

Fig. 2.8 Variation of the dimensionless bubble radius under external acoustic field. (**a**) Single-frequency. (**b**) Dual-frequency. Reprinted with the permission from Ref. [26] Copyright (2017) (ELSEVIER)

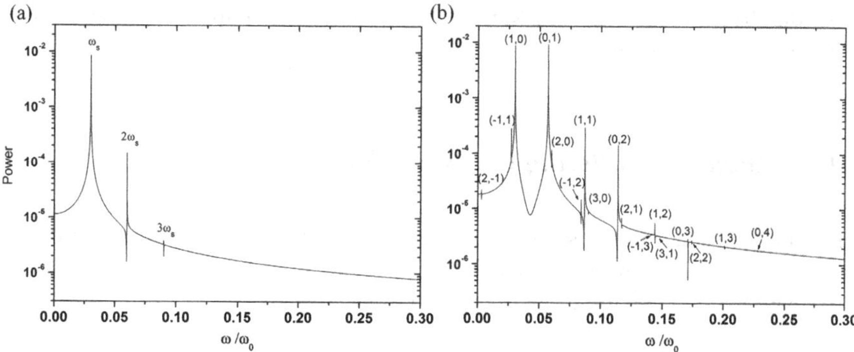

Fig. 2.9 Power spectrum of bubble oscillations under the external acoustic field. (**a**) Single-frequency. (**b**) Dual-frequency. Reprinted with the permission from Ref. [26] Copyright (2017) (ELSEVIER)

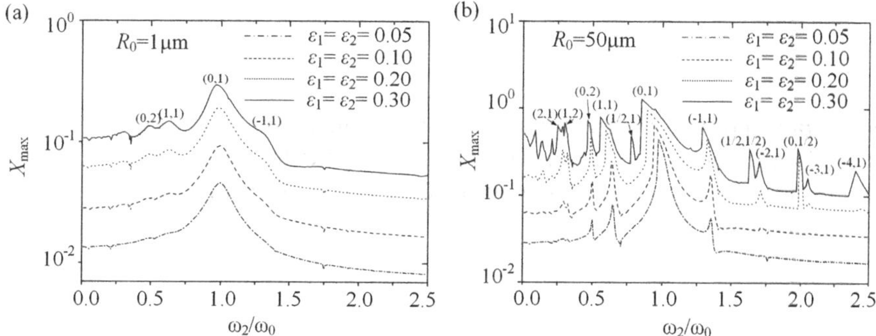

Fig. 2.10 Effect of bubble radius and acoustic pressure amplitude on the response curve of bubble oscillations. (**a**) $R_0 = 1\,\mu m$. (**b**) $R_0 = 50\,\mu m$. (Reprinted with the permission from Ref. [26] Copyright (2017) (ELSEVIER))

harmonic bands are labeled $(a, 0)$ and $(0, b)$, and the combination resonance band is labeled as (a, b) for which neither a nor b is zero.

Figure 2.10 demonstrates the effect of bubble radius and acoustic pressure amplitude on the response curve of bubble oscillations. Figure 2.10a, b have bubble equilibrium radii of 1 μm and 50 μm, respectively. $X_{max} = (R_{max} - R_0)/R_0$, where R_{max} is the maximum radius of the bubble during steady-state oscillations. Both the equilibrium radius of the bubble and the acoustic pressure amplitude affect the bubble oscillations. For small bubbles (Fig. 2.10a), bubble oscillations are suppressed, leading to strong suppression of harmonic and combination resonances. At the same pressure amplitude, the higher order harmonics and the combination resonance are more pronounced for large size bubbles (Fig. 2.10b). The amplitudes of the main resonance, harmonics and subharmonics increase as the acoustic pressure amplitude increases. The higher-order combination resonances (i.e., large $|a| + |b|$ values) have smaller amplitudes than the lower-order combination resonances, but grow more rapidly with increasing pressure amplitude. The widths of all resonances increase with pressure amplitude.

2.4.2 Acoustic Scattering Cross Section of Bubbles

The acoustic scattering cross section is used to evaluate the scattering ability of bubbles when excited by an acoustic wave. The radiation pressure $p_{rad}(r, t)$ at a radial coordinate r from the bubble center is given by [27].

$$p_{rad}\left(r,t\right) = \frac{\rho_1 R}{r}\left(2\dot{R}^2 + R\ddot{R}\right)$$
(2.165)

with the amplitude of radiation pressure [27].

$$\left|p_{rad}\right| = \left|\frac{B}{r}\right|$$
(2.166)

where B/r represents the amplitude of the divergent spherical scattered wave produced by the bubble oscillation. The acoustical scattering cross section (σ_s) of an oscillating bubble in a liquid is defined as the square of the ratio of the scattered wave to the incident wave [28].

$$\sigma_s = 4\pi \left|\frac{B}{A}\right|^2$$
(2.167)

where A is the amplitude of the incident wave.

For a dual-frequency acoustic field, the acoustic pressure $p_s(t)$ in Eqs. (2.43) is given by [28].

$$p_s\left(t\right) = p_0\left[1 + \varepsilon_1 \cos\left(2\pi f_1 t\right) + \varepsilon_2 \cos\left(2\pi f_2 t\right)\right]$$
(2.168)

where f_1 and f_2 are the frequencies of the acoustic fields with angular frequencies ω_1 and ω_2, respectively.

Assuming that under the dual-frequency acoustic field, the solution to the Keller equation [Eqs. (2.43)–(2.46)] is [28].

$$\frac{R}{R_0} = 1 + \varepsilon_1 x$$
(2.169)

Then the solution for the bubble motion equation becomes [28].

$$x = A_{11} \cos\left(\omega_1 t + \delta_{11}\right) + A_{12} \cos\left(\omega_2 t + \delta_{12}\right)$$
(2.170)

where the expressions for A_{11}, A_{12}, δ_{11} and δ_{12} are provided in Eqs. (2.95)–(2.98). Substituting Eq. (2.169) into Eq. (2.165) and retaining terms up to first order in ε_1, we obtain [28].

$$B = \rho_1 R_0^3 \varepsilon_1 \left(A_{11}^2 \omega_1^4 + A_{12}^2 \omega_2^4 \right)^{\frac{1}{2}} \tag{2.171}$$

Therefore, the analytical solution for the acoustic scattering cross section under the dual-frequency excitation is [28].

$$\sigma_s = 4\pi \left| \frac{B}{A} \right|^2 = 4\pi \frac{\left(\rho_1 R_0^3 / p_0 \right)^2 \left(A_{11}^2 \omega_1^4 + A_{12}^2 \omega_2^4 \right)}{1 + \left(p_{A2} / p_{A1} \right)^2} \tag{2.172}$$

For single-frequency acoustic excitation, the analytical solution for the acoustic scattering cross section is [28].

$$\sigma_s = \frac{4\pi R_0^2}{\left(\dfrac{\omega_0^2}{\omega_s^2} - 1 \right)^2 M^2 + \left(\dfrac{4\mu_1}{\omega_s \rho_1 R_0^2} + \dfrac{R_0}{\omega_s c_1} M \omega_0^2 \right)^2} C_1 \tag{2.173}$$

with

$$C_1 = \frac{\left[\sin\left(\dfrac{2\pi R_0}{\lambda_1} \right) / \left(\dfrac{2\pi R_0}{\lambda_1} \right) \right]^2}{1 + \left(\dfrac{2\pi R_0}{\lambda_1} \right)^2} \tag{2.174}$$

where λ_1 is the wavelength of acoustic wave in the liquid.

The acoustic scattering cross section can also be solved by numerical simulation. In numerical simulations, the bubble motion equations [Eqs. (2.43)–(2.46) and (2.168)] are first solved using the explicit Runge-Kutta method, followed by the calculation of the radiation pressure and acoustic scattering cross section via Eqs. (2.165)–(2.167).

Figure 2.11 shows the bubble radius and radiation pressure under dual-frequency acoustic excitation. Under dual-frequency excitation, the bubble radius and radiation pressure exhibit complex variations and are periodic. Figure 2.12 shows the acoustic scattering cross-section of the bubble under dual-frequency excitation numerical simulation and analytical solution. The following constants were used for both numerical simulations and analytical solutions: $\rho_1 = 998.20$ kg/m^2, $\mu_1 = 1.0$ mPa $\cdot$ s, $\sigma = 0.0728$ N/m, $c_1 = 1486$ m/s, $p_0 = 101300$ Pa, $\kappa = 1.33$. The acoustical scattering cross section was evaluated at $r = 1 \times 10^{-3}$ m from the origin. For dual-frequency acoustic excitation, the total input power $P_e = \sqrt{p_{A1}^2 + p_{A2}^2}$ was

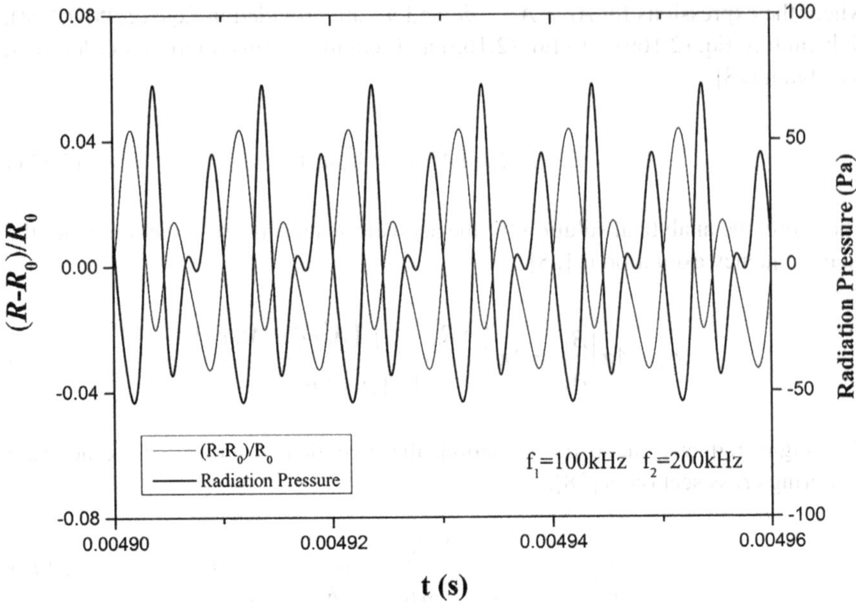

Fig. 2.11 Instantaneous bubble radius and corresponding radiation pressure under dual-frequency acoustic excitation. (Reprinted with the permission from Ref. [28] Copyright (2017) (ELSEVIER))

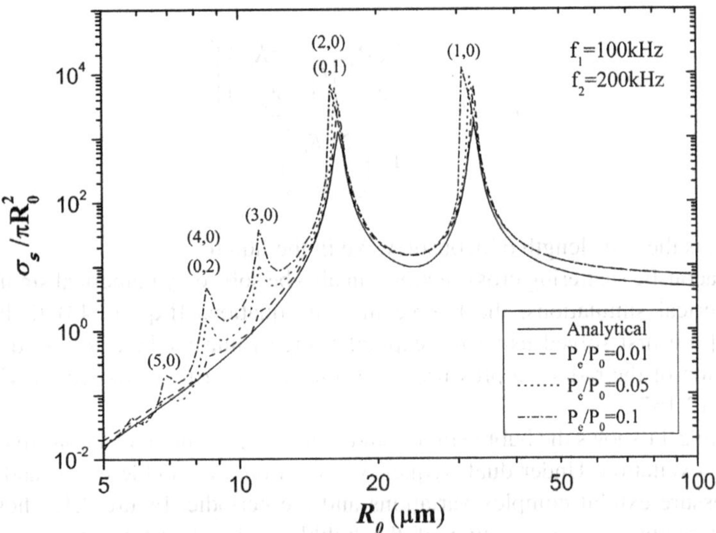

Fig. 2.12 Acoustical scattering cross section against equilibrium bubble radius under dual-frequency acoustic excitation. (Reprinted with the permission from Ref. [28] Copyright (2017) (ELSEVIER))

held constant to enable comparison, where p_{A1} and p_{A2} are the amplitudes of the two acoustic waves, respectively. The pressure amplitude ratio of the two acoustic waves is defined as $N = p_{A2}/p_{A1}$. The labels (a, b) in Fig. 2.11 correspond to those in Fig. 2.9. The scattering cross section demonstrates nonlinear behavior, as revealed by the numerical simulations. According to Eq. (2.172), the analytical solution is independent of the acoustic wave amplitude. At low pressure amplitudes, such as $P_e/P_0 = 0.01$, the analytical solution aligns closely with the numerical results. However, as P_e the numerically simulated scattering cross section exhibits pronounced nonlinearity, including the generation of harmonics. When $P_e/P_0 = 0.1$, higher-order harmonics become evident.

Figure 2.13 illustrates the effect of the pressure amplitude ratio N on the acoustic scattering cross section of the bubble. The curves labeled f_1 and f_2 represent single-frequency excitations at frequencies f_1 and f_2, respectively. Under dual-frequency excitation, the σ_s curves exhibit additional resonance regions and greater amplitudes compared to the single-frequency excitations. The parameter N governs the energy distribution between the two frequency components and significantly influences the scattering cross section. When $N < 1$, as N decreases, the peak of the main resonance (0,1) becomes smaller and narrower, while the peak of the main resonance (1,0) is almost identical to that of the single-frequency excitation at frequency f_1, since more energy is allocated to the acoustic wave with amplitude p_{A1}. When $N > 1$, as N increases, the peak of the main resonance (1,0) becomes smaller and narrower, while the peak of the main resonance (0,1) is almost identical to that of the single-frequency excitation at frequency f_2, since more energy is allocated to the acoustic

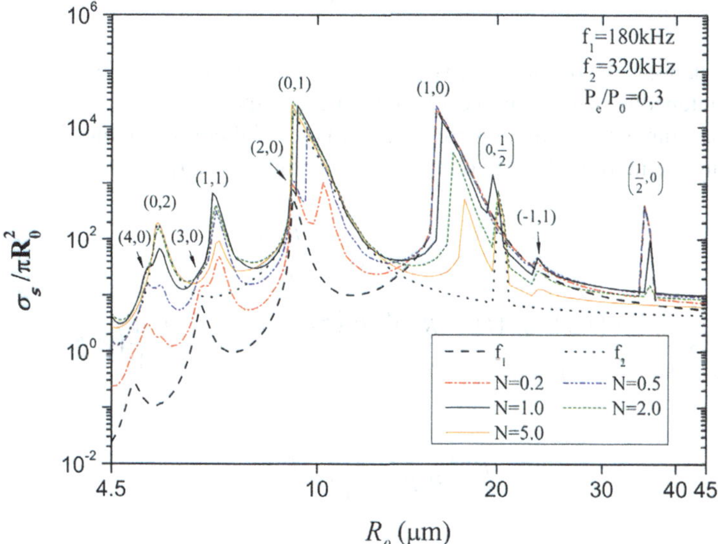

Fig. 2.13 Effect of the pressure amplitude ratio N of the two acoustic waves on the acoustic scattering cross-section of the bubble. (Reprinted with the permission from Ref. [28] Copyright (2017) (ELSEVIER))

wave with amplitude p_{A2}. For subharmonics, greater energy allocation to one frequency enhances the peaks of both the main resonance and subharmonics tied to that frequency. For combination resonances [i.e., $(1, 1)$ and $(-1, 1)$], the scattering cross section reaches its maximum value when $N = 1$.

2.4.3 Secondary Bjerknes Force

When bubbles oscillate in an acoustic field, the radiation pressure produced by neighboring bubbles can induce mutual attraction or repulsion among them, a phenomenon termed the "secondary Bjerknes force." This force results in the translational motion of the bubbles, influencing their tendency to aggregate or disperse, which is essential for comprehending the dynamics of bubble clouds and their practical applications. This section considers the secondary Bjerknes force between two oscillating bubbles (labeled as "bubble 1" and "bubble 2") in a liquid under dual-frequency acoustic excitation.

The radiation pressure produced by bubble 2 at the center of bubble 1 (p_{rad1}) or by bubble 1 at the center of bubble 2 (p_{rad2}) is given by [29].

$$p_{rad1} = \frac{\rho_1}{L} \frac{\mathrm{d}}{\mathrm{d}t} \left(\dot{R}_2 R_2^2 \right) \tag{2.175}$$

$$p_{rad2} = \frac{\rho_1}{L} \frac{\mathrm{d}}{\mathrm{d}t} \left(\dot{R}_1 R_1^2 \right) \tag{2.176}$$

where L denotes the distance between the centers of the two bubbles. R_1 and R_2 are the instantaneous radii of bubble 1 and bubble 2, respectively.

Accounting for the radiation pressure from the neighboring bubble, the equations governing the motion of the two interacting bubbles are [29].

$$\left(1 - \frac{\dot{R}_1}{c_1} \right) R_1 \ddot{R}_1 + \frac{3}{2} \left(1 - \frac{\dot{R}_1}{3c_1} \right)$$
$$= \left(1 + \frac{\dot{R}_1}{c_1} \right) \frac{p_1(R,t) - p_s(t)}{\rho_1} + \frac{R_1}{\rho_1 c_1} \frac{\mathrm{d}\left[p_1(R,t) - p_s(t) \right]}{\mathrm{d}t} - \frac{1}{L} \frac{\mathrm{d}}{\mathrm{d}t} \left(\dot{R}_2 R_2^2 \right) \tag{2.177}$$

$$\left(1 - \frac{\dot{R}_2}{c_1} \right) R_2 \ddot{R}_2 + \frac{3}{2} \left(1 - \frac{\dot{R}_2}{3c_1} \right)$$
$$= \left(1 + \frac{\dot{R}_2}{c_1} \right) \frac{p_2(R,t) - p_s(t)}{\rho_1} + \frac{R_2}{\rho_1 c_1} \frac{\mathrm{d}\left[p_2(R,t) - p_s(t) \right]}{\mathrm{d}t} - \frac{1}{L} \frac{\mathrm{d}}{\mathrm{d}t} \left(\dot{R}_1 R_1^2 \right) \tag{2.178}$$

with

$$p_1\left(R_1,t\right) = \left(p_0 + \frac{2\sigma}{R_{01}}\right)\left(\frac{R_{01}}{R_1}\right)^{3\kappa} - \frac{2\sigma}{R_1} - \frac{4\left(\mu_1 + \mu_{\text{th}}\right)\dot{R}_1}{R_1} \tag{2.179}$$

$$p_2\left(R_2,t\right) = \left(p_0 + \frac{2\sigma}{R_{02}}\right)\left(\frac{R_{02}}{R_2}\right)^{3\kappa} - \frac{2\sigma}{R_2} - \frac{4\left(\mu_1 + \mu_{\text{th}}\right)\dot{R}_2}{R_2} \tag{2.180}$$

$$p_s\left(t\right) = p_0\left(1 + \varepsilon_1 e^{i\omega_1 t} + \varepsilon_2 e^{i\omega_2 t}\right) \tag{2.181}$$

where R_{01} and R_{02} are the equilibrium bubble radii of bubbles 1 and 2, respectively.

Given that the secondary Bjerknes force arises from the mutual interaction between two bubbles, we focus on the force exerted on bubble 1 by bubble 2. When the bubble size is small relative to the wavelength of the acoustic wave emitted by the other bubble, the translational force on bubble 1 is approximated as the negative gradient of the radiation pressure multiplied by the volume of bubble 1. Thus, the secondary Bjerknes force F_B produced by bubble 2 on bubble 1 is defined as the time average of the translational force [30].

$$F_B = -\left\langle v_1 \nabla p_2 \right\rangle \tag{2.182}$$

with

$$v_1 = \frac{4}{3}\pi R_1^3 \tag{2.183}$$

$$\nabla p_2 = \frac{\rho_1}{L^2}\frac{\text{d}}{\text{d}t}\left(\dot{R}_2 R_2^2\right)e_{12} \tag{2.184}$$

where v_1 represents the instantaneous volume of bubble 1, ∇p_2 is the pressure gradient induced by bubble 2 at the center of bubble 1, and e_{12} is the unit vector pointing from bubble 1 to bubble 2.

From Eqs. (2.182)–(2.184), the secondary Bjerknes force F_B can be determined once the instantaneous radii of bubble 1 and bubble 2 are known. An analytical solution for F_B under dual-frequency acoustic excitation is obtained by applying perturbation methods to solve Eqs. (2.177)–(2.178). Alternatively, a numerical solution is computed by solving these equations [Eqs. (2.177)–(2.178)] using the explicit Runge-Kutta method.

The analytical solution for the secondary Bjerknes force is introduced as follows. Assume that the solutions to Eqs. (2.177)–(2.178) are [30].

$$R_1 = R_{01}\left(1 + x_1\right) \tag{2.185}$$

$$R_2 = R_{02}\left(1 + x_2\right) \tag{2.186}$$

Consequently, the inhomogeneous equations for the harmonic oscillators representing the two bubbles are [30].

$$\ddot{x}_1 + 2\beta_1 \dot{x}_1 + \omega_{01}^2 x_1 + \frac{R_{02}^2}{R_{01}^2}\xi_2\,\ddot{x}_2 = -\frac{\alpha_1}{M_1}\left(N_{11}\varepsilon_1 e^{i\omega_1 t} + N_{12}\varepsilon_2 e^{i\omega_2 t}\right) \tag{2.187}$$

$$\ddot{x}_2 + 2\beta_2 \dot{x}_2 + \omega_{02}^2 x_2 + \frac{R_{01}^2}{R_{02}^2}\xi_1\,\ddot{x}_1 = -\frac{\alpha_2}{M_2}\left(N_{21}\varepsilon_1 e^{i\omega_1 t} + N_{22}\varepsilon_2 e^{i\omega_2 t}\right) \tag{2.188}$$

where

$$\omega_{0j}^2 = \frac{1}{M_j \rho_1 R_{0j}^2}\left[3\kappa\left(p_0 + \frac{2\sigma}{R_{0j}}\right) - \frac{2\sigma}{R_{0j}}\right] \tag{2.189}$$

$$\beta_j = \beta_{vj} + \beta_{thj} + \beta_{acj} = \frac{2\left(\mu_1 + \mu_{th}\right)}{\rho_1 R_{0j} M_j} + \frac{R_{0j}}{2c_1}\omega_{0j}^2 \tag{2.190}$$

$$\xi_j = \frac{R_{0j}}{L} \tag{2.191}$$

$$\alpha_j = \frac{p_0}{\rho_1 R_{0j}^2} \tag{2.192}$$

$$M_j = 1 + \frac{R_{0j}}{c_1}\frac{4\mu_1}{\rho_1 R_{0j}^2} \tag{2.193}$$

$$N_{jk} = 1 + \frac{\omega_k R_{0j}}{c_1}i \tag{2.194}$$

Here, $j = 1, 2$, $k = 1, 2$. ω_{0j} is the natural frequency of the bubble j. β_{vj}, β_{thj}, and β_{acj} are the total, viscous, thermal and acoustic damping constants of the bubble j, respectively.

Given that only linear oscillations are considered (up to first order in ε), the solutions for x_1 and x_2 are [30].

$$x_1 = A_{11}e^{i\omega_1 t} + A_{12}e^{i\omega_2 t} \tag{2.195}$$

$$x_2 = A_{21}e^{i\omega_1 t} + A_{22}e^{i\omega_2 t} \tag{2.196}$$

Substituting Eqs. (2.195) and (2.196) into Eqs. (2.187) and (2.188) and solving, we obtain [30].

$$A_{11} = -\frac{\alpha_1 \varepsilon_1}{X_1}\left[\frac{N_{21}}{M_2}\xi_2\omega_1^2 + \frac{N_{11}}{M_1}\left(\omega_{02}^2 - \omega_1^2 + 2\beta_2\omega_1 i\right)\right] \qquad (2.197)$$

$$A_{21} = -\frac{\alpha_2 \varepsilon_1}{X_1}\left[\frac{N_{11}}{M_1}\xi_1\omega_1^2 + \frac{N_{21}}{M_2}\left(\omega_{01}^2 - \omega_1^2 + 2\beta_1\omega_1 i\right)\right] \qquad (2.198)$$

$$A_{12} = -\frac{\alpha_1 \varepsilon_2}{X_2}\left[\frac{N_{22}}{M_2}\xi_2\omega_2^2 + \frac{N_{12}}{M_1}\left(\omega_{02}^2 - \omega_2^2 + 2\beta_2\omega_2 i\right)\right] \qquad (2.199)$$

$$A_{22} = -\frac{\alpha_2 \varepsilon_2}{X_2}\left[\frac{N_{12}}{M_1}\xi_1\omega_2^2 + \frac{N_{22}}{M_2}\left(\omega_{01}^2 - \omega_2^2 + 2\beta_1\omega_2 i\right)\right] \qquad (2.200)$$

with

$$X_1 = \left(\omega_{01}^2 - \omega_1^2 + 2\beta_1\omega_1 i\right)\left(\omega_{02}^2 - \omega_1^2 + 2\beta_2\omega_1 i\right) - \xi_1\xi_2\omega_1^4 \qquad (2.201)$$

$$X_2 = \left(\omega_{01}^2 - \omega_2^2 + 2\beta_1\omega_2 i\right)\left(\omega_{02}^2 - \omega_2^2 + 2\beta_2\omega_2 i\right) - \xi_1\xi_2\omega_2^4 \qquad (2.202)$$

Following Doinikov's framework [31], define $v_1 = \mathrm{Im}\,(V_1)$ and $\nabla p_2 = \mathrm{Im}\,(\nabla P_2)$. For the explicit cases, Eqs. (2.183) and (2.184) become [30].

$$V_1 = V_C + V_{11} + V_{12} \approx \frac{4}{3}\pi R_{01}^3\left(1 + 3A_{11}e^{i\omega_1 t} + 3A_{12}e^{i\omega_2 t}\right) \qquad (2.203)$$

$$\nabla p_2 = \nabla P_{21} + \nabla P_{22} \approx \frac{\rho_1}{L^2}R_{02}^3\, x_2\, \boldsymbol{e}_{12}$$

$$= -\frac{\rho_1}{L^2}R_{02}^3\left(\omega_1^2 A_{21}e^{i\omega_1 t} + \omega_2^2\right)A_{22}e^{i\omega_2 t}\boldsymbol{e}_{12} \qquad (2.204)$$

Since V_C is a constant, $\langle V_C \nabla p_2\rangle = 0$ does not contribute to the secondary Bjerknes force. After time-averaging, terms $\langle e^{2i\omega_1 t}\rangle$ and $\langle e^{2i\omega_2 t}\rangle$ are both zero. Therefore, the secondary Bjerknes force is expressed as [30].

$$F_B = \frac{1}{2}\mathrm{Re}\left(\bar{V}_{11}\nabla P_{21} + \bar{V}_{12}\nabla P_{22}\right) \qquad (2.205)$$

where, a bar over the symbol denotes the complex conjugate.

Substituting Eqs. (2.203) and (2.204) into Eq. (2.205), the analytical solution for the secondary Bjerknes force under dual-frequency acoustic excitation is [30].

$$F_B = 2\pi\rho_1 R_{01}^2 R_{02}^2 \xi_1\xi_2\left[\omega_1^2\,\mathrm{Re}\left(\bar{A}_{11}A_{12}\right) + \omega_2^2\,\mathrm{Re}\left(\bar{A}_{12}A_{22}\right)\right] \qquad (2.206)$$

For numerical simulation, combining Eqs. (2.182)–(2.184), one can obtain [30].

$$F_B = -\langle v_1 \nabla p_2 \rangle = -\left\langle \frac{v_1 \rho_1}{L^2} \frac{d}{dt}\left(\dot{R}_2 R_2^2\right) \right\rangle e_{12} = -\left\langle \frac{v_1 \rho_1}{4\pi L^2} \frac{d^2 v_2}{dt^2} \right\rangle e_{12} \qquad (2.207)$$

Integrating Eq. (2.207) over a period of the volume oscillations by partial integration, the secondary Bjerknes force is [30].

$$F_B = \frac{\rho_1}{4\pi L^2} \langle \dot{v}_1 \dot{v}_2 \rangle e_{12} \qquad (2.208)$$

with

$$\dot{v}_1 = 4\pi R_1^2 \dot{R}_1 \qquad (2.209)$$

$$\dot{v}_2 = 4\pi R_2^2 \dot{R}_2 \qquad (2.210)$$

Solving Eqs. (2.177)–(2.178) with the explicit Runge-Kutta formula and applying Eqs. (2.208)–(2.210) yields the numerical secondary Bjerknes force. From Eq. (2.208), the distance L between the two bubbles does not affect the direction of the secondary Bjerknes force, so the result is expressed using the secondary Bjerknes force coefficient [29].

$$f_B = \frac{\rho_1}{4\pi} \langle \dot{v}_1 \dot{v}_2 \rangle \qquad (2.111)$$

If $f_B > 0$, the two bubbles attract each other; if $f_B < 0$, the two bubbles repel each other.

Figure 2.14 shows a comparison of the analytical and numerical solutions for f_B under dual-frequency acoustic excitation as a function of R_{02}. Results for four different values of P_e/P_0 are presented. R_{r1} and R_{r2} denote the resonance radii corresponding to the driving frequencies. When $R_{02} < R_{r2}$, as R_{02} increases, a positive f_B indicates that the two bubbles attract each other. As R_{02} approaches R_{r2} and continues to increase, f_B reaches a peak and then sharply decreases to negative values, indicating a transition from attraction to repulsion between the bubbles. When $R_{r2} < R_{02} < R_{r1}$, f_B changes from negative to positive as R_{02} increases, meaning the bubbles switch from repulsion to attraction. As R_{02} approaches R_{r1} and further increases, f_B again reaches a positive peak and subsequently decreases to negative values. When $R_{02} > R_{r1}$, f_B remains negative, indicating mutual repulsion between the bubbles. For small pressure amplitudes, as shown in Fig. 2.14a, b, the analytical and numerical solutions match well geometrically. As the pressure amplitude increases, the peak value of f_B increases. The peak of f_B in the numerical solution is lower than that in the analytical solution and shifts towards smaller bubble radii. In Fig. 2.14d, the numerical solution predicts a peak in f_B at the resonance radius corresponding to

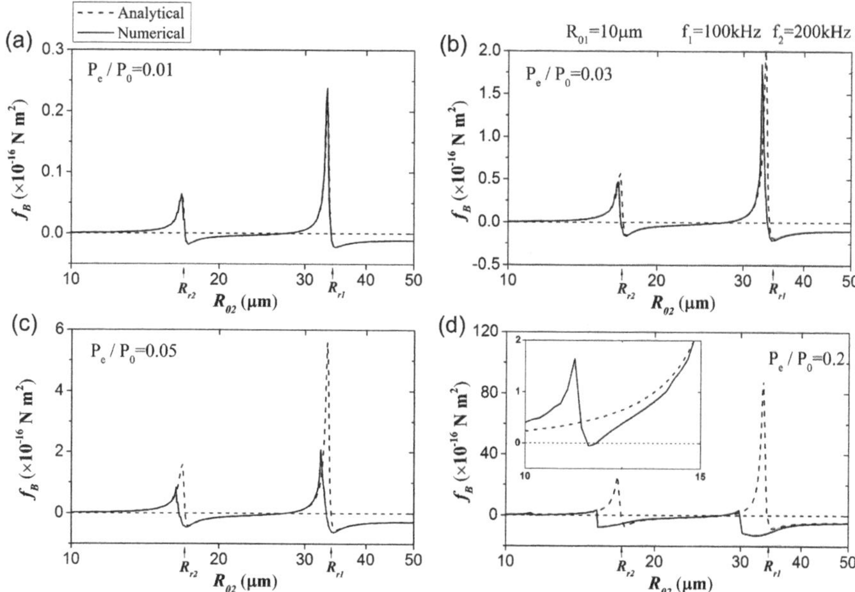

Fig. 2.14 Comparison of the analytical and numerical solutions for f_B under dual-frequency acoustic excitation as a function of R_{02}. (Reprinted with the permission from Ref. [30] Copyright (2016) (ELSEVIER))

$f_1 + f_2$, which the analytical solution fails to predict. Therefore, the analytical solution is applicable for low pressure amplitudes.

Figure 2.15 displays the variation of f_B in the $R_{01} - R_{02}$ plane under single-frequency and dual-frequency acoustic excitation. Figures 2.15a, b correspond to the frequencies of $f_1 = 100$ kHz and $f_2 = 200$ kHz, respectively. Figure 2.15c shows the result for the dual-frequency acoustic excitation. The red regions correspond to $f_B < 0$, indicating that the secondary Bjerknes force is repulsive. The gray regions correspond to $f_B > 0$, indicating that the secondary Bjerknes force is attractive. Under single-frequency excitation, the $R_{01} - R_{02}$ plane is divided into four regions, with the "boundaries" between regions corresponding to the resonance radius at the driving frequency. The values of f_B are large near these boundaries. Under dual-frequency excitation, the $R_{01} - R_{02}$ plane is divided into nine regions. These nine regions can be classified into three types: repulsive regions, attractive regions, and uncertain regions. The uncertain regions are where f_B may be positive or negative, covering the ranges $R_{01} < R_{r2} < R_{02} < R_{r1}$, $R_{02} < R_{r2} < R_{01} < R_{r1}$, $R_{r2} < R_{02} < R_{r1} < R_{01}$, and $R_{r2} < R_{01} < R_{r1} < R_{02}$. According to Eq. (2.206), when P_e is limited, the value of f_B under dual-frequency excitation can be regarded as a linear combination of f_B under the two single-frequency excitations. In the uncertain regions, by adding a second acoustic excitation, f_B can be forced or suppressed, leading to a change in its sign, which depends on the relative values of f_B corresponding to the two frequency components.

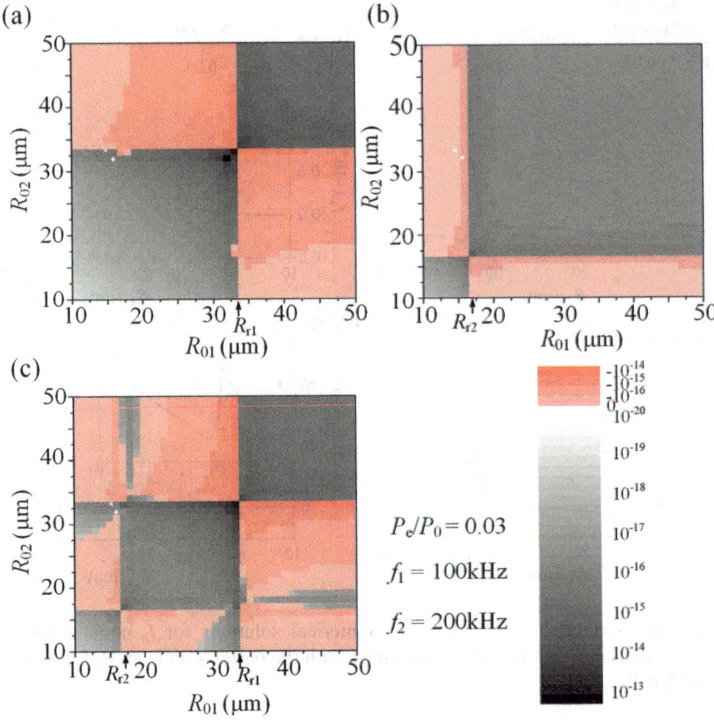

Fig. 2.15 Variation of f_B in the $R_{01} - R_{02}$ plane under single-frequency and dual-frequency acoustic excitation. (Reprinted with the permission from Ref. [30] Copyright (2016) (ELSEVIER))

References

1. Prosperetti A, Lezzi A. Bubble dynamics in a compressible liquid. Part 1. First-order theory. J Fluid Mech. 1986;168:457–78.
2. Fujikawa S, Akamatsu T. Effects of the non-equilibrium condensation of vapour on the pressure wave produced by the collapse of a bubble in a liquid. J Fluid Mech. 1980;97(3):481–512.
3. Lezzi A, Prosperetti A. Bubble dynamics in a compressible liquid. Part 2. Second-order theory. J Fluid Mech. 1987;185:289–321.
4. Jamshidi R, Brenner G. Dissipation of ultrasonic wave propagation in bubbly liquids considering the effect of compressibility to the first order of acoustical Mach number. Ultrasonics. 2013;53(4):842–8.
5. Blake FG. The onset of cavitation in liquids; technical memo no.12. Cambridge, MA: Acoustics Research Laboratory, Harvard University; 1949.
6. Epstein P, Plesset M. On the stability of gas bubbles in liquid-gas solutions. J Chem Phys. 1950;18(11):1505–9.
7. Hsieh D, Plesset M. Theory of rectified diffusion of mass into gas bubbles. J Acoust Soc Am. 1961;33(2):206–15.
8. Eller A, Flynn H. Rectified diffusion during nonlinear pulsations of cavitation bubbles. J Acoust Soc Am. 1965;37(3):493–503.

9. Crum L, Hansen G. Generalized equations for rectified diffusion. J Acoust Soc Am. 1982;72(5):1586–92.
10. Zhang Y, Li S. A general approach for rectified mass diffusion of gas bubbles in liquids under acoustic excitation. J Heat Transf. 2014;136(4):042001.
11. Zhang Y, Li S. Mass transfer during radial oscillations of gas bubbles in viscoelastic mediums under acoustic excitation. Int J Heat Mass Transf. 2014;69:106–16.
12. Keller J, Miksis M. Bubble oscillations of large amplitude. J Acoust Soc Am. 1980;68(2):628–33.
13. Crum L, Mao Y. Acoustically enhanced bubble growth at low frequencies and its implications for human diver and marine mammal safety. J Acoust Soc Am. 1996;99(5):2898–907.
14. Zhang Y. Rectified mass diffusion of gas bubbles in liquids under acoustic field with dual frequencies. Int Commun Heat Mass Transfer. 2012;39(10):1496–9.
15. Zhang Y. Analysis of radial oscillations of gas bubbles in Newtonian or viscoelastic mediums under acoustic excitation. University of Warwick; 2012.
16. Prosperetti A. Thermal effects and damping mechanisms in the forced radial oscillations of gas bubbles in liquids. J Acoust Soc Am. 1977;61(1):17–27.
17. Keller J, Kolodner I. Damping of underwater explosion bubble oscillations. J Appl Phys. 1956;27(10):1152–61.
18. Khabeev NS. Resonance properties of vapor bubbles. J Appl Math Mech. 1981;45(4):512–7.
19. Shima A. The natural frequency of a bubble oscillating in a viscous compressible liquid. J Basic Eng. 1970;92(3):555–61.
20. Zhang Y. Heat transfer across interfaces of oscillating gas bubbles in liquids under acoustic excitation. Int Commun Heat Mass Transfer. 2013;43:1–7.
21. Zhang Y, Li S. Notes on radial oscillations of gas bubbles in liquids: thermal effects. J Acoust Soc Am. 2010;128(5):EL306-EL309.
22. Devin C. Survey of thermal, radiation, and viscous damping of pulsating air bubbles in water. J Acoust Soc Am. 1959;31(12):1654–67.
23. Prosperetti A. Bubble phenomena in sound fields: part one. Ultrasonics. 1984;22(2):69–77.
24. Zhang Y, Li S. Effects of liquid compressibility on radial oscillations of gas bubbles in liquids. J Hydrodyn. 2012;24(5):760–6.
25. Dormand PP. A family of embedded Runge-Kutta formulae. J Comput Appl Math. 1980;6(1):19–26.
26. Zhang Y, Zhang Y, Li S. Combination and simultaneous resonances of gas bubbles oscillating in liquids under dual-frequency acoustic excitation. Ultrason Sonochem. 2017;35:431–9.
27. Yang X, Church C. A model for the dynamics of gas bubbles in soft tissue. J Acoust Soc Am. 2005;118(6):3595–606.
28. Zhang Y, Li S. Acoustical scattering cross section of gas bubbles under dual-frequency acoustic excitation. Ultrason Sonochem. 2015;26:437–44.
29. Mettin R, Akhatov I, Parlitz U, et al. Bjerknes forces between small cavitation bubbles in a strong acoustic field. Phys Rev E. 1997;56(3):2924–31.
30. Zhang Y, Zhang Y, Li S. The secondary Bjerknes force between two gas bubbles under dual-frequency acoustic excitation. Ultrason Sonochem. 2016;29:129–45.
31. Doinikov AA. Effects of the second harmonic on the secondary Bjerknes force. Phys Rev E. 1999;59(3):3016–21.

Chapter 3
Propagation of Sound Waves in Vapor/Gas/Liquid Multiphase Flow

This chapter focuses on the propagation of acoustic wave in liquid containing vapor-gas bubbles. First, four typical acoustic wave propagation models are presented and the effective region is described in detail. Second, the process of solving for wave speed is elaborated. Then, the types and typical characteristics of waves are explained. Finally, the effects of vapor mass fraction, void fraction, and mixed bubble radius on the critical frequency, constant wave speed, and minimum wave speed are specifically analyzed.

3.1 Prediction Model of Wave Speed

When cavitation occurs, a large number of vapor bubbles are generated in the liquid. As the non-condensable gases inside the liquid diffuse into the interface of the cavitation bubbles, vapor-gas mixture bubbles are formed, thus forming a vapor-gas-liquid multiphase flow. The presence of mixture bubbles causes a significant increase in the compressibility of the liquid, which corresponds to a change in the physical properties of the propagation medium. Therefore, it is of great theoretical significance to study the propagation of acoustic waves in vapor-gas-liquid mixed flows. In addition, in many engineering applications, non-condensable gases are introduced into liquids in order to increase the chemical reaction rate and optimize the reactor structure. Acoustic wave propagation in vapor-gas mixed bubbly liquids is the theoretical basis for the study of many problems (e.g., sonochemistry [1], medical [2], ocean engineering [3]), so the study of acoustic wave propagation in vapor-gas mixed bubbly liquids is of great practical significance in engineering.

The investigation of acoustic wave propagation in liquid containing vapor-gas mixed bubbly liquids can be traced back to the mid-twentieth century. With the advancement of numerical simulation methods (e.g., boundary element method [4] and finite volume method [5]) and experimental techniques (e.g., high-speed

© The Author(s), under exclusive license to Springer Nature
Switzerland AG 2025

J. Hu et al., *Multiphase Flow with Bubbles*, SpringerBriefs in Energy,
https://doi.org/10.1007/978-3-031-99216-2_3

photography [6]), Acoustic wave propagation models in vapor-gas mixed bubbly liquids are continuously improved. For bubbly flows with large void fraction, Wood Model [7], Brennen Model [8] and Crespo Model [9] were proposed. For bubbly flows with small void fraction, Ando Model [10] and Prosperetti Model [11] were proposed. These models are described below. In addition, the effective region of the acoustic propagation model in the full frequency range and the influencing factors are described.

3.1.1   Derivation of the Acoustic Wave Propagation Model

The models of acoustic wave propagation in vapor-gas mixed bubbly liquids are based on a number of assumptions. The assumptions are listed below [12]:

(a) Liquid mixtures containing non-condensable gases and vapors are considered to be homogeneous fluids,
(b) Mixed bubbles are considered as a sphere,
(c) Acoustic waves are a single frequency,
(d) Bubbles in mixed liquids are mono-disperse.

In thermodynamic theory, the acoustic wave speed (c_m) can be expressed as [12].

$$c_m = \left(\frac{dp}{d\rho_m} \right)^{\frac{1}{2}} \tag{3.1}$$

where p and ρ_m are the pressure and average density of the mixture, respectively, $\rho_m = \beta\rho_g + (1 - \beta)\rho_l$ with ρ_g and ρ_l denote the densities of the gas bubble and the liquid, respectively. β denotes the volume fraction of gas per unit volume of mixture and is defined as void fraction.

The equilibrium pressure (p_G) inside the bubble is [12].

$$p_G = p_0 + \frac{2\sigma}{R_0} \tag{3.2}$$

where p_0 represents the pressure of the environmental fluid, σ represents the surface tension, R_0 is the equilibrium mixed bubble radius.

The acoustic wave speed in a mixed fluid can be expressed as [13].

$$\frac{1}{c_m^2} = \left(\frac{d\rho_m}{dp} \right) = (1-\beta)\frac{d\rho_l}{dp} + \beta\frac{d\rho_g}{dp} + \left(\rho_g - \rho_l \right)\frac{d\beta}{dp} \tag{3.3}$$

The speed of acoustic wave in pure liquid and pure gas are labeled c_l and c_g, respectively. The formulas for c_l and c_g, are [12].

$$c_l^2 = \frac{\mathrm{d}p}{\mathrm{d}\rho_l} \tag{3.4}$$

$$c_g^2 = \frac{\mathrm{d}p}{\mathrm{d}\rho_g} \tag{3.5}$$

Since $(\rho_g c_g)^2 \ll (\rho_l c_l)2$, Eq. (3.3) can be simplified to [13].

$$\frac{1}{c_m^2} = \frac{(1-\beta)^2}{c_l^2} + \frac{\beta^2}{c_g^2} + \frac{\rho_l \beta (1-\beta)}{p} \tag{3.6}$$

where, in the mixed fluid, the value of β is between 0 and 1. The first and second terms on the right side of Eq. (3.6) can be ignored. From the Eq. (3.2), the Wood model [7] was derived as

$$c_m^2 = \frac{p_G}{\rho_l \beta (1-\beta)} \tag{3.7}$$

The Wood model is related to the mixed bubble radius and is independent of the frequency of acoustic waves. This equation is widely used to calculate the propagation of acoustic wave speed in uniformly mixed bubbly fluids.

Crespo [9] introduced coefficients α and κ based on the Wood model. α is a coefficient that depends on the viscous length, thermal diffusion length, and the relative size of the mixed bubble radius. κ is a coefficient that reflects the thermodynamic state inside the bubble, with the value of κ ranging from 1 to γ. γ represents the ratio of specific heat capacities. $\kappa = 1$ indicates an isothermal state, while $\kappa = \gamma$ indicates an adiabatic state. For detailed calculations regarding κ, readers can refer to Zhang [14]. The Crespo model [9] is

$$c_m^2 = \frac{\alpha \kappa p_G}{\rho_l \beta (1-\beta)} \tag{3.8}$$

Brennen [8] proposed that the acoustic impedance of a mixture fluid is obtained through a weighted average based on the volume fractions of each component. Therefore, for gas-liquid mixed bubbly fluids, the formula for acoustic impedance is [8].

$$\frac{1}{\rho_m c_m^2} = \frac{\beta}{\kappa p_G} + \frac{1-\beta}{\rho_l c_l^2} \tag{3.9}$$

where $\kappa p_G \ll \rho_l c_l^2$, the second term on the right side of Eq. (3.9) can be ignored [12]. The Brennen model [8] can be expressed as

$$\frac{1}{c_m^2} = \frac{\beta}{\kappa p_G} \left[\rho_1 (1-\beta) + \rho_g \beta \right] \tag{3.10}$$

For bubbles in bubbly flow with a low void fraction, it behaves as isothermal under low frequencies, i.e., $\kappa = 1$. The dispersion relation is [11].

$$\frac{c_1^2}{c_m^2} = 1 + \frac{\beta \rho_1 c_1^2}{p_G} \frac{1}{1 - \dfrac{2\sigma}{3R_0 p_G}} \tag{3.11}$$

Ignoring the surface tension, the formula for the Ando model [10] is

$$c_m^2 = \frac{c_1^2}{1 + \dfrac{\beta \rho_1 c_1^2}{p_G}} \tag{3.12}$$

where, in most cases, $\beta \rho_1 c_1^2 \gg 1$. Therefore, the Prosperetti [11] model obtained from the simplified form (3.12) is

$$c_m = \frac{p_0}{\beta \rho_1} \tag{3.13}$$

3.1.2   *The Effective Area of the Acoustic Wave Propagation Model*

The propagation of acoustic waves in a gas-liquid mixed bubbly liquids has two important effects: the viscous effect and the thermal diffusion effect. A discussion of the viscous length δ_v, thermal diffusion length δ_{th}, and the mixed bubble radius R_0 indicates the impact of these two effects on the acoustic wave speed.

The formulas for the viscous length δ_v and thermal diffusion length δ_{th} are [12].

$$\delta_v = \sqrt{\frac{v}{\omega}} \tag{3.14}$$

$$\delta_{th} = \sqrt{\frac{D_g}{\omega}} \tag{3.15}$$

where, v and ω are the kinematic viscosity of the liquid and the frequency of the external excitation in the liquid, respectively. D_g is the thermal diffusion rate of the gas inside the bubble.

At different frequencies, the magnitudes of the aforementioned three characteristic lengths vary, and thus the effects of viscosity and thermal diffusion on acoustic waves are also different. In addition, these three lengths divide the full frequency range into three regions, and the acoustic wave speed formulas for these three regions need to be slightly modified. The following will introduce these.

For low frequencies, the relationship between the sizes of the three feature lengths is [12].

$$R_0 \ll \delta_v \ll \delta_{th} \tag{3.16}$$

For middle frequencies, the relationship between the sizes of the three feature lengths is [12].

$$\delta_v \ll R_0 \ll \delta_{th} \tag{3.17}$$

For high frequencies, the relationship between the sizes of the three feature lengths is [12].

$$\delta_v \ll \delta_{th} \ll R_0 \tag{3.18}$$

Eq. (3.16) indicates that at low frequencies, viscosity limits the velocity difference between the bubble and the liquid, resulting in no relative motion between the bubble and the liquid. Furthermore, δ_{th} is much greater than R_0, which means that heat diffusion within the bubble occurs rapidly, and the temperature of the bubble interface is close to that of the surrounding liquid. Therefore, the bubble can be approximated as an isothermal bubble. In other words, in this region, α and κ can be treated as unity in this region [12]. At this time, the propagation of acoustic wave speed is consistent with the Wood model. At middle frequencies, δ_v is much smaller than R_0, the effect of viscosity is reduced, and a speed difference occurs between the bubbles and the liquid, resulting in relative motion. Meanwhile, $\alpha = [1 + 2\beta(1 - \beta)/\Gamma]$, Γ is the coefficient of the force acting on the bubble [9], its value is 1. The propagation of acoustic waves at this moment behaves differently from the Wood model, whose formula is [9].

$$c_m^2 = \frac{\left[1 + 2\beta(1 - \beta)p_G\right]}{\rho_l \beta(1 - \beta)} \tag{3.19}$$

At high frequencies, the relative motion between the bubbles and the liquid is enhanced, while the heat transfer between the two is neglected. The bubbles can be approximated as adiabatic bubbles. At this time, $\kappa = \gamma$, $\alpha = [1 + 2\beta(1 - \beta)/\Gamma]$, the formula for acoustic wave propagation is [9].

$$c_m^2 = \frac{\left[1 + 2\beta(1 - \beta)\gamma p_G\right]}{\rho_l \beta(1 - \beta)} \tag{3.20}$$

where, in most engineering applications, the volumetric fraction of gas is much less than 1. Therefore, β^2 in Eqs. (3.19) and (3.20) can be neglected, and the simplified formulas are shown separately as follows [9].

$$c_{\mathrm{m}}^2 = \frac{(1+2\beta)\,p_{\mathrm{G}}}{\rho_{\mathrm{l}}\beta(1-\beta)} \tag{3.21}$$

$$c_{\mathrm{m}}^2 = \frac{(1+2\beta)\gamma\,p_{\mathrm{G}}}{\rho_{\mathrm{l}}\beta(1-\beta)} \tag{3.22}$$

Figure 3.1 exhibits the variation of three characteristic lengths with frequency when the mixed bubble radius is 100 μm. The changes of δ_{th} and δ_{v} are similar, both first decrease rapidly and then slowly lower as the frequency increases. In particular, the value of δ_{th} is always higher than that of δ_{v}. R_0 is independent of frequency, and its value remains unchanged. In the figure, both curves of R_0 with δ_{v} and δ_{th} have intersection points ($\omega_1 = 85.7$ rad/s, $\omega_2 = 1900$ rad/s). These two intersection points divide the entire parameter space into low frequency (Region 1), middle frequency (Region 2), and high frequency (Region 3). The acoustic wave speed formulas for these three regions correspond to Eqs. (3.7), (3.19), and (3.20) respectively. The acoustic wave speed is greatly related to the mixed bubble radius and void fraction.

Figure 3.2 shows the effect of void fraction β on the three acoustic wave propagation models represented by Eqs. (3.7), (3.19), and (3.20). As the void fraction increases,

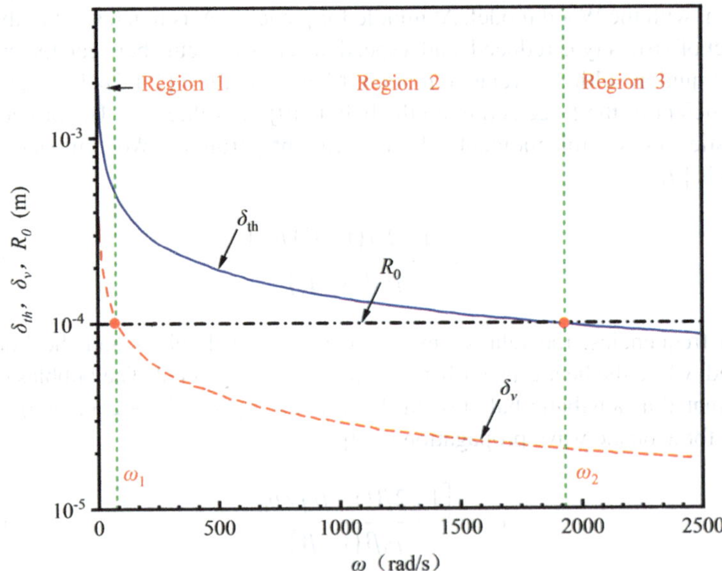

Fig. 3.1 The variations of mixed bubble radius, viscous length and thermal diffusion length versus frequency. $R_0 = 100$ μm. (Reprinted with the permission from Ref. [12] Copyright (2017) (IOP Publishing Ltd))

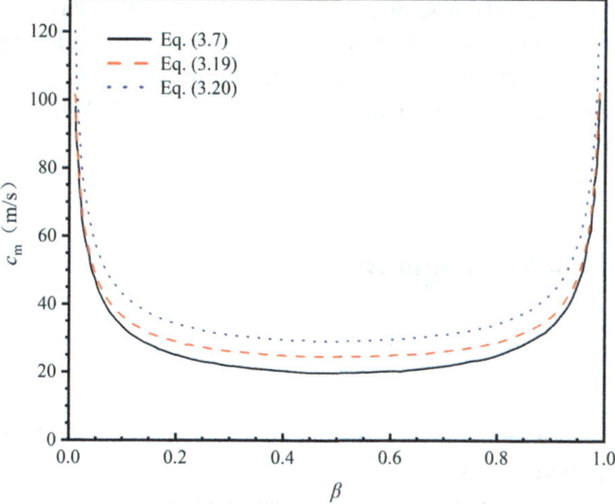

Fig. 3.2 The variations of the acoustic wave speed predicted by Eqs. (3.7), (3.19) and (3.20) versus the void fraction. $R_0 = 100$ μm. (Reprinted with the permission from Ref. [12] Copyright (2017) (IOP Publishing Ltd))

the variation of acoustic wave speed in the three acoustic wave propagation models is basically consistent. When $\beta < 0.3$, the acoustic wave speed decreases rapidly at first with an increase in β, and then decreases slowly; when $\beta > 0.7$, the acoustic wave speed increases slowly at first with an increase in β, and then increases rapidly; when $0.3 \leq \beta \leq 0.7$, the acoustic wave speed changes very little with an increase in β.

Figure 3.3 compares the effectiveness of the Wood model [Eq. (3.7)], Brennen model [Eq. (3.10)], Ando model [Eq. (3.12)], and Prosperetti model [Eq. (3.13)]

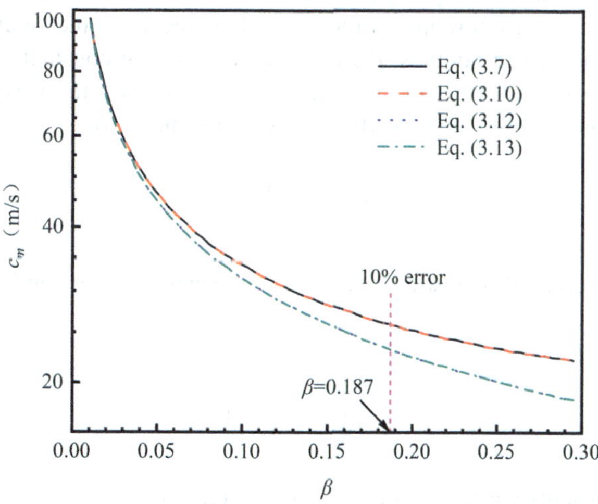

Fig. 3.3 Comparisons of simplified formulae of acoustic wave speed. Formulas are Eqs. (3.7), (3.10), (3.12) and (3.13). $T_0 = 300$ K, $R_0 = 100$ μm. (Reprinted with the permission from Ref. [12] Copyright (2017) (IOP Publishing Ltd))

with $\beta = 0.3$. The variations in acoustic wave speed for the Wood model and Brennen model are identical, while the variations for the Ando model and Prosperetti model are also identical. As β increases, the prediction error of the Ando model and Prosperetti model increases compared to the Wood model. When $\beta = 0.187$, the error reaches 10%.

3.2 Solution of Wave Speed

This section provides the process for calculating wave speed. The assumptions are listed below [15]:

(a) Both water vapor and non-condensable gases exist inside the bubble,
(b) Bubbles are spherical and monodispersed,
(c) The distance between the bubbles is far enough to ignore the interactions between them,
(d) The amplitude of the acoustic wave is limited, and there is no chaotic oscillation occurring,
(e) Ignore the impact of changes in environmental temperature.

The formula for the complex acoustic wave speed through a mono-disperse bubbly flow is [13].

$$\frac{1}{c_m^2} = \frac{1}{c_1^2} + \frac{4\pi n R_0}{\omega_0^2 - \omega^2 + 2i\beta_{\text{tot}}\omega} \tag{3.23}$$

where c_m is the phase speed of the composite wave; c_1 is the wave speed in pure liquid; n is the number of vapor-gas mixed bubbles in a unit volume of the mixed flow; R_0 is the equilibrium radius of the mixed bubbles; ω_0 is the natural frequency of the oscillating mixed bubbles; β_{tot} is the total damping constant, which causes attenuation of wave velocity during the propagation of acoustic waves. Therefore, the void fraction β can be given by the following formula [16]:

$$\beta = \frac{4\pi n R_0^3}{3} \tag{3.24}$$

The phase speed c_{ph} of acoustic waves in a vapor-gas mixture can be represented as [13].

$$c_{\text{ph}} = \left(\text{Re}\left(\frac{1}{c_m} \right) \right)^{-1} \tag{3.25}$$

where, Re represents the real part of a complex function.

For the closed Eq. (3.23), the expressions for β_{tot} and ω_0 are shown as follows [17]:

$$\beta_{\mathrm{tot}} = \frac{2\left(\mu_1 + \mu_{\mathrm{th}}\right)\varphi}{M\rho_1 R_0^{\,2}} + \frac{R_0}{2c_1}\tilde{\omega}_0^{\,2} - \frac{1}{2}\mathrm{Re}\left(J_0^*\right)\frac{\rho_b}{\rho_1}\omega \tag{3.26}$$

$$\omega_0^{\,2} = \frac{p_{\mathrm{in,eq}}}{M\rho_1 R_0^{\,2}}\left(3\kappa - \frac{2\sigma}{p_{\mathrm{in,eq}} R_0}\right) + Im\left(J_0^*\right)\frac{\rho_b}{\rho_1}\omega^2 \tag{3.27}$$

with

$$\kappa = \frac{1}{3}\mathrm{Re}\left(\Phi\right) \tag{3.28}$$

$$\mu_{\mathrm{th}} = \frac{4p_{\mathrm{in,eq}}}{\omega}Im\left(\Phi\right) \tag{3.29}$$

$$M = 1 + \frac{4\left(\mu_1 + \mu_{\mathrm{th}}\right)}{\rho_1 R_0^{\,2}}\frac{R_0}{c_1} \tag{3.30}$$

$$\varphi = 1 - \frac{\rho_b}{\rho_1}Im\left(J_0^*\right) \tag{3.31}$$

$$\tilde{\omega}_0^{\,2} = \frac{p_{\mathrm{in,eq}}}{M\rho_1 R_0^{\,2}}\left(3\kappa - \frac{2\sigma}{p_{\mathrm{in,eq}} R_0}\right) \tag{3.32}$$

$$\alpha_0 = \frac{p_0}{\rho_1 R_0^{\,2}} \tag{3.33}$$

$$\varepsilon = \frac{p_A}{p_0} \tag{3.34}$$

where J_0^* is a dimensionless characteristic mass transfer flux, ε is the dimensionless amplitude of the external acoustic excitation, ρ_b is the density of the bubbles, Φ is the transfer function between the pressure inside the bubbles and the oscillation of the mixed bubble radius. The total damping constant is divided into four parts, which are viscous damping constant β_{vis}, thermal damping constant β_{th}, acoustic damping constant β_{ac}, and mass transfer damping constant β_{mass}. For β_{mass}, all terms related to mass transfer are included in this damping mechanism (e.g., the J_0^* term). $p_{\mathrm{in,eq}}$ is the equilibrium pressure in the bubble, p_A is the amplitude of external acoustic excitation, p_0 is the ambient pressure.

For the values of J_0^* and Φ in Eqs. (3.26) and (3.27), the complete solution obtainable from the calculations by Fuster and Montel [18] is referenced.

$$J_0^* = J_c^*\left(1 - \Delta H_{\mathrm{vap}}^* \Delta T_c^I\right) \tag{3.35}$$

$$\Phi = 3\gamma / \left\{ 1 - 3(\gamma-1)iPe_b^{-1}\left[\sqrt{Pe_b i}\coth\left(\sqrt{Pe_b i}\right)-1\right]\left(1-\Delta T_c'\right)\right.$$
$$\left. + 3\gamma U_c^*\left(\Delta H_{vap}^* \Delta T_c' - 1\right)\right\} \tag{3.36}$$

with

$$Pe_b = \frac{\omega R_0^2}{D_b^T} \tag{3.37}$$

$$\Delta T_c' = \frac{k_b}{k_l}\left[\sqrt{Pe_b i}\coth\left(\sqrt{Pe_b i}\right)-1+J_c^* Pe_b \Delta H_{vap}^*\left(\frac{\gamma}{\gamma-1}\right)\right] /$$
$$\left\{1+\sqrt{Pe_l i}+\frac{k_b}{k_l}\left[\sqrt{Pe_b i}\coth\left(\sqrt{Pe_b i}\right)-1+J_c^* Pe_b\left(\Delta H_{vap}^*\right)^2\left(\frac{\gamma}{\gamma-1}\right)\right]\right\} \tag{3.38}$$

$$\Delta H_{vap}^* = \frac{\Delta H_{vap}}{c_{p,b}T_0} \tag{3.39}$$

$$J_c^* = \frac{J_{max}^*}{1+J_{max}^*\left(J_{c,eq}^*\right)^{-1}} \tag{3.40}$$

where

$$J_{c,eq}^* = \frac{\sqrt{iSh_D}\coth\left(\sqrt{SS_D}\right)-1}{Sh_D}\frac{Y_0}{1-Y_0} \tag{3.41}$$

$$Sh_D = \frac{\omega R_0^2}{D_{w/b}} \tag{3.42}$$

$$J_{max}^* = \alpha_{evap}\frac{P_{in,eq}}{\rho_{b,0}R_0\omega\sqrt{2\pi r_w T_0}} \tag{3.43}$$

$$Pe_l = \frac{\omega R_0^2}{D_l^T} \tag{3.44}$$

where, $\Delta T_c'$ is the dimensionless temperature change at the bubble interface, γ is the specific heat ratio, D_b^T is the thermal diffusion coefficient of the bubble, ΔH_{vap}^* is the vaporization enthalpy of the liquid, $c_{p,b}$ is the specific heat inside the bubble, T_0 is the ambient temperature, k_b and k_l are the thermal conductivity coefficients of the bubble and the liquid, respectively, $D_{w/b}$ is the diffusion coefficient of water vapor inside the bubble, α_{evap} is the permeability coefficient, r_w is the gas constant for vapor, D_l^T is the thermal diffusion coefficient of the liquid, $Y_0 = 0.99$ [18].

It is worth noting that the presence of J_0^* indicates that the case of mass transfer has been considered. If mass transfer is ignored (i.e., $J_0^* = J_c^* = 0$) and heat transfer ($\varnothing T_c^I = 0$) are ignored, Eqs. (3.26), (3.27), and (3.35) can be respectively simplified to [17].

$$\beta_{\text{tot}} = \beta_{\text{vis}} + \beta_{\text{th}} + \beta_{\text{ac}} = \frac{2\mu_1}{M\rho_1 R_0^2} + \frac{2\mu_{\text{th}}}{M\rho_1 R_0^2} + \frac{R_0}{2c_1}\omega_0^2 \tag{3.45}$$

$$\omega_0^2 = \frac{P_{\text{in,eq}}}{M\rho_1 R_0^2}\left(3\kappa - \frac{2\sigma}{P_{\text{in,eq}} R_0}\right) \tag{3.46}$$

$$\Phi = \frac{3\gamma}{1 - 3(\gamma-1)iPe_b^{-1}\left[\sqrt{Pe_b i}\,\coth\left(\sqrt{Pe_b i}\right) - 1\right]} \tag{3.47}$$

3.3 Characteristics of Acoustic Waves

3.3.1 Partitioning of Acoustic Waves

As can be seen from Sect. 3.1, the propagation of acoustic waves in gas-liquid mixed bubbly liquids is strongly dependent on frequency. Section 3.2 presents the process for solving wave speed under general conditions. This section mainly discusses the simplified formulas for acoustic waves in the high-frequency region and low-frequency region.

In the low-frequency region, the mixed bubbles will oscillate at far below resonance, i.e., $\omega \ll \omega_0$. In this case, the damping term can be neglected. The complex wave speed c_m is equal to the phase speed c_{ph}. Therefore, by substituting n in Eq. (3.23) with the simplified wave speed formula derived from Eq. (3.24) is [16].

$$\frac{1}{c_m^2} = \frac{1}{c_1^2} + \frac{3\beta}{R_0^2}\frac{1}{\omega_0^2 - \omega^2} \tag{3.48}$$

where the c_1 can be ignored. Since $\omega \ll \omega_0$, the wave speed at this time can be considered as the low frequency wave speed c_L, and its formula is [16].

$$c_L = \sqrt{\frac{\omega_0^2 R_0^2}{3\beta}} \tag{3.49}$$

As the frequency decreases, the $\text{Re}(\Phi)$ term decreases, and the natural frequency value in Eq. (3.49) also decreases accordingly. Therefore, in the low-frequency region, the wave speed will approach the value of c_L from a higher value. To close Eq. (3.29), it is necessary to obtain the low frequency limit of the transfer function

Φ. In the low frequency case, the Sherwood number and the bubble Peclet number are very low, therefore, the temperature of the interface can be ignored. Based on the above assumptions, Eq. (3.36) can be simplified to [18].

$$\Phi \approx 3(1-Y_0) \tag{3.50}$$

Therefore, the inherent frequency of the oscillating mixed bubble is [16].

$$\omega_0^{\,2} = \frac{P_{in,eq}}{\rho_1 R_0^{\,2}}\left[3(1-Y_0) - \frac{2\sigma}{P_{in,eq}R_0}\right] \tag{3.51}$$

From the simultaneous Eqs. (3.49) and (3.50), we can obtain c_L is [16].

$$c_L = \sqrt{\frac{1}{3\beta\rho_1}\left[3(1-Y_0) - \frac{2\sigma}{P_{in,eq}R_0}\right]} \tag{3.52}$$

In order to demonstrate the effect of surface tension on c_L, a coefficient ψ is defined as [16].

$$\psi = \sqrt{1 + \frac{2\sigma}{p_0 R_0}\left[1 - \frac{1}{3(1-Y_0)}\right]} \tag{3.53}$$

Therefore, ψ can be transformed into [16].

$$c_L = \psi\sqrt{\frac{(1-Y_0)p_0}{\beta\rho_1}} \tag{3.54}$$

From Eq. (3.54), it can be seen that the surface tension term only becomes significant when the vapor mass fraction is high. For large bubbles or low vapor mass fraction, surface tension term will varnish and $\psi = 1$. According to the research by Fuster and Montel [18], in the case where both the Sherwood number and the bubble Peclet number are small (e.g. less than 1) and the interfacial temperature changes are neglected, the expression for c_L is [16].

$$c_L = \frac{c_1}{\sqrt{1 + \frac{c_1^{\,2}\beta\rho_1}{(1-Y_0)p_0}}} \tag{3.55}$$

where $\dfrac{c_1^{\,2}\beta\rho_1}{(1-Y_0)p_0} \ll 1$, Eq. (3.55) can be simplified to [16].

$$c_L = \sqrt{\frac{(1-Y_0)p_0}{\beta\rho_1}} \tag{3.56}$$

A parameter corresponding to the low frequency wave speed is the low frequency critical frequency f_L. The wave speed below the critical frequency is considered constant. Before deriving f_L, we first assume that the wave speed corresponding to f_L is $k_L c_L$, where $k_L \approx 1$, and neglect the c_1^2 term. Eq. (3.48) can be simplified to [16].

$$\frac{1}{k_L^2 c_L^2} = \frac{3\beta}{R_0^2} \frac{1}{\omega_0^2 - (2\pi f_L)^2}$$

(3.57)

By combining Eqs. (3.52) and (3.53), we can obtain [16].

$$f_L = \frac{1}{2\pi\sqrt{\rho_1 R_0}} \left[3(1-Y_0) p_{\text{in,eq}} - \frac{2\sigma}{R_0} - 3(1-Y_0) k_L^2 \psi p_0 \right]$$

(3.58)

Eq. (3.58) shows that f_L is related to the mixed bubble radius and the vapor mass fraction, but is independent of the void fraction.

In the high-frequency range, the mixed bubbles will oscillate above the resonance point, that is, $\omega \gg \omega_0$. The damping term is also neglected. The wave speed formula is [16].

$$\frac{1}{c_m^2} = \frac{1}{c_1^2} - \frac{3\beta}{R_0^2} \frac{1}{\omega_0^2 - \omega^2}$$

(3.59)

where the wave speed is the high-frequency wave speed c_H, and the high frequency limit is $c_H = c_L$. Corresponding to the high-frequency wave speed is a high frequency f_H, above which the wave speed is a constant. Unlike c_L, c_H is a constant in the high-frequency range that is independent of the mixed bubble radius, void fraction, and vapor mass fraction. In Eq. (3.29), $\text{Re}(\Phi) = 3\gamma$ [14], therefore, ω_0 is also a definite value. From Eq. (3.59), it can be seen that as the frequency increases, the wave speed decreases until it reaches the minimum value c_H.

The high frequency f_H is derived from Eq. (3.59). Since $\omega \gg \omega_0$, the second term on the right side of Eq. (3.59) is very small, comparing the first and second terms on the right side yields a coefficient k_H [16].

$$k_H = \frac{3\beta}{R_0^2} \frac{c_1^2}{(2\pi f_L)^2 - \omega_0^2}$$

(3.60)

where $k_H \ll 1$.

The inherent frequency of the oscillating mixed bubble is [14].

$$\omega_0^2 = \frac{p_{\text{in,eq}}}{\rho_1 R_0^2} \left(3\gamma - \frac{2\sigma}{p_{\text{in,eq}} R_0} \right)$$

(3.61)

By solving Eqs. (3.60) and (3.61) simultaneously, the high frequency f_H can be obtained as [16].

$$f_\mathrm{H} = \frac{1}{2\pi R_0}\left[\frac{3\beta c_1^2}{k_\mathrm{H}} + \frac{P_\mathrm{in,eq}}{\rho_1 R_0^{\,2}}\left(3\gamma - \frac{2\sigma}{p_\mathrm{in,eq} R_0}\right)\right]^{\frac{1}{2}} \qquad (3.62)$$

From Eq. (3.62), it can be seen that f_H is only related to void fraction and mixed bubble radius, and is not dependent on vapor mass fraction. If surface tension is ignored, it can be simplified to [16].

$$f_\mathrm{H} = \frac{1}{2\pi R_0}\left(\frac{3\beta c_1^2}{k_\mathrm{H}} + \frac{3\gamma p_0}{\rho_1}\right)^{\frac{1}{2}} \qquad (3.63)$$

Figure 3.4 illustrates the variation of wave speed with frequency. The range of frequency variation is from 1×10^{-4} Hz to 5×10^7 Hz, which basically covers all frequency ranges in engineering applications. In Fig. 3.4, vapor mass fraction Y_0 is 0.8, the void fraction β is 0.10%, and the mixed bubble radius R_0 is 100 μm. From Fig. 3.4, the full frequency can be divided into three regions based on the variation of wave speed, namely low frequency region, middle frequency region, and high-frequency region. The critical frequencies of these three regions correspond to the low frequency f_L and the high frequency f_H. In the low frequency region ($\omega \ll \omega_0$), as the frequency increases, the wave speed remains constant, i.e., c_L. The wave speed in the low frequency region is independent of frequency. In the middle frequency region ($\omega \approx \omega_0$), as the frequency increases, the wave speed oscillates

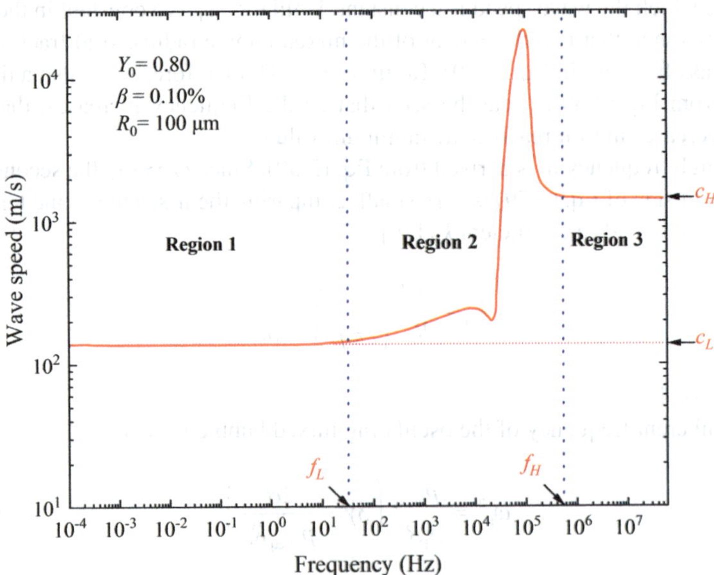

Fig. 3.4 The variation of wave speed with frequency. $Y_0 = 0.80$. $\beta = 0.10\%$. $R_0 = 100$ μm. (Reprinted with the permission from Ref. [16] Copyright (2018) (ELSEVIER))

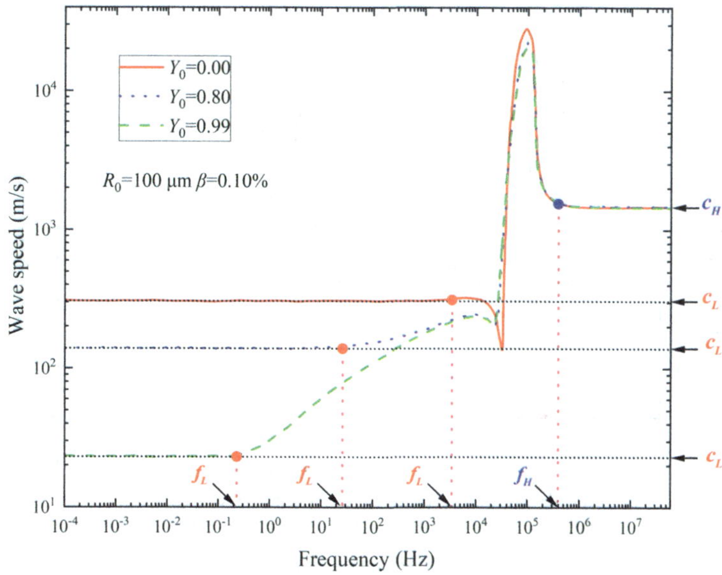

Fig. 3.5 The effect of the vapor mass fraction Y_0 on the speed of acoustic wave. $\beta = 0.10\%$, $R_0 = 100$ μm. (Reprinted with the permission from Ref. [16] Copyright (2018) (ELSEVIER))

dramatically, first experiencing a slow fluctuation, then a sharp fluctuation. In the high-frequency region ($\omega \gg \omega_0$), as the frequency increases, the wave speed also remains constant, i.e., c_H. In this region, the wave speed is also independent of frequency.

Figure 3.5 shows the effect of the vapor mass fraction Y_0 on the speed of acoustic wave. The three curves represent the propagation of acoustic waves in gas, vapor-gas mixture bubbles, and nearly vapor. f_L and c_L gradually decrease with the increase of Y_0, while f_H and c_H remain constant. Fig. 3.5 indicates that f_L and c_L are sensitive to vapor mass fraction, while f_H and c_H are insensitive to Y_0.

Figure 3.6 shows the effect of the void fraction β on acoustic wave speed. The acoustic wave speed curves vary similarly with frequency at different porosities. As β increases, f_L remains constant while c_L gradually decreases. f_H increases with β, while c_H remains at a constant value. Figure 3.6 indicates that f_H and c_L are sensitive to porosity, while f_L and c_H are not.

Figure 3.7 shows the effect of the mixed bubble radius on acoustic wave speed. As R_0 increases, f_L gradually increases while f_H decreases, with c_L and c_H remaining constant. Fig. 3.7 indicates that f_L and f_H are functions of mixed bubble radius, while c_L and c_H are not. The effects of these three parameters on cutoff frequency and constant wave speed will be analyzed in Sect. 3.4 and 3.5.

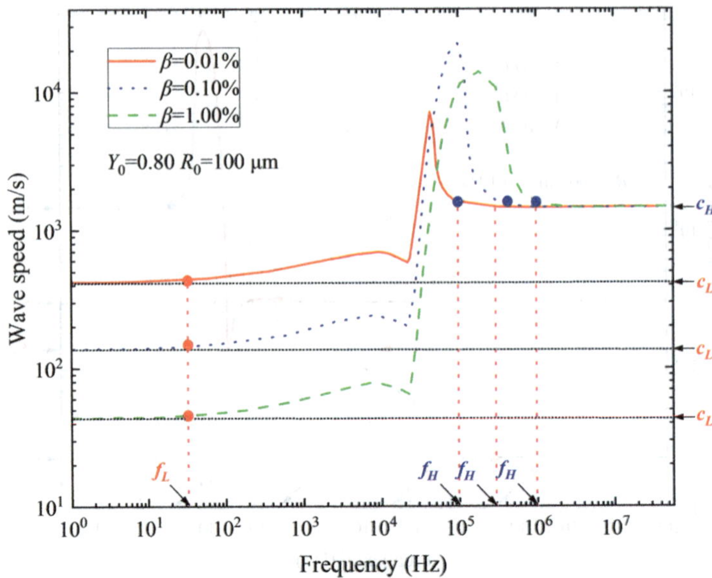

Fig. 3.6 The effect of the void fraction β on the speed of acoustic wave. $Y_0 = 0.80$. $R_0 = 100$ μm. (Reprinted with the permission from Ref. [16] Copyright (2018) (ELSEVIER))

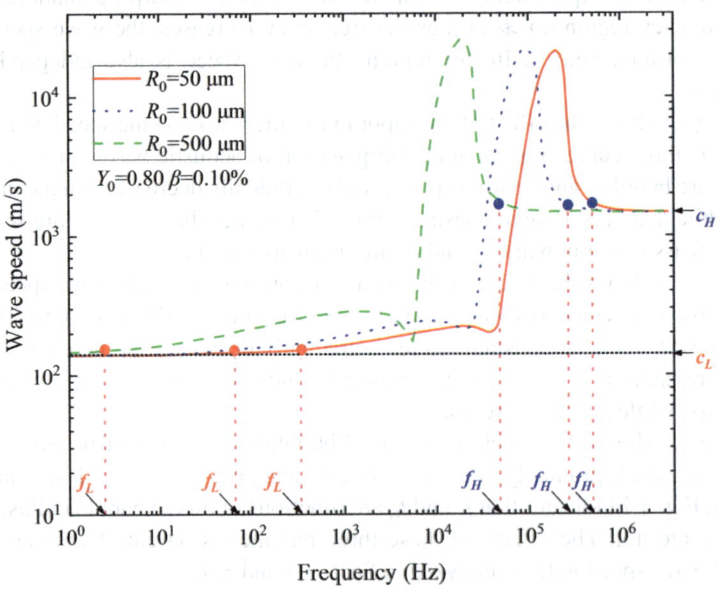

Fig. 3.7 The effect of the void fraction on the speed of acoustic wave. $Y_0 = 0.80$. $\beta = 0.10\%$. (Reprinted with the permission from Ref. [16] Copyright (2018) (ELSEVIER))

3.3.2   Types of Acoustic Wave Propagation

The minimum wave speed C_{min} is of great significance in the design of industrial ultrasonic chemical reactors [19]. Fig. 3.8 illustrates the propagation and attenuation of acoustic waves in bubbles, vapor bubbles, and gas-liquid mixed bubbles as a function of frequency. In Fig. 3.8a, for the case of gas bubbles, the wave speed is approximately constant in the low-frequency range, then oscillates around the middle frequency range, and finally reaches a constant value in the high-frequency range (i.e., the wave speed in pure liquid is 1500 m/s). For the case of vapor bubbles, the wave speed is very low in the low-frequency range and increases with frequency, with a greater range of change compared to the other two cases. For the case of vapor-gas mixed bubbles, the wave speed changes are intermediate between those of gas bubbles and vapor bubbles. As shown in Fig. 3.8b, in the low-frequency range, wave attenuation increases with the increase of vapor content, as the damping effect of vapor bubbles is very significant [17].

In the three cases mentioned above, the position of C_{min} can be divided into two situations: one is the C_{min} close to the middle frequency region, and the other is the C_{min} located in the low frequency region. Based on the different positions of C_{min}, the wave propagation types can be categorized into two types. Figure 3.9 shows the variation of the minimum wave speed C_{min} with the vapor mass fraction Y_0, under the condition of a fixed void fraction and mixed bubble radius. As Y_0 increases, C_{min} initially shows an approximately linear increasing trend, and then rapidly decreases after reaching the critical vapor mass fraction Y_{cr}. C_{min} on both sides of Y_{cr} belong to type 1 and type 2 [19].

Figure 3.10 demonstrates the effect of mixed bubble radius and void fraction on Y_0. It can be seen from the figure that for a constant mixed bubble radius, Y_{cr} increases with increasing void fraction. However, for high void fraction, the curves of Y_{cr} almost coincide, which indicates that the effect of void fraction on Y_{cr} is very

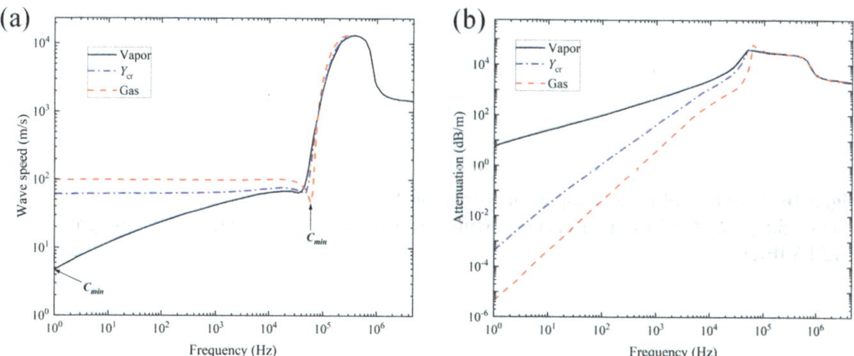

Fig. 3.8 The propagation (**a**) and attenuation (**b**) of acoustic waves in gas, vapor, and vapor-gas mixed bubbles as a function of frequency. $\beta = 1.0\%$. $R_0 = 50$ μm. (Reprinted with the permission from Ref. [19] Copyright (2018) (ELSEVIER))

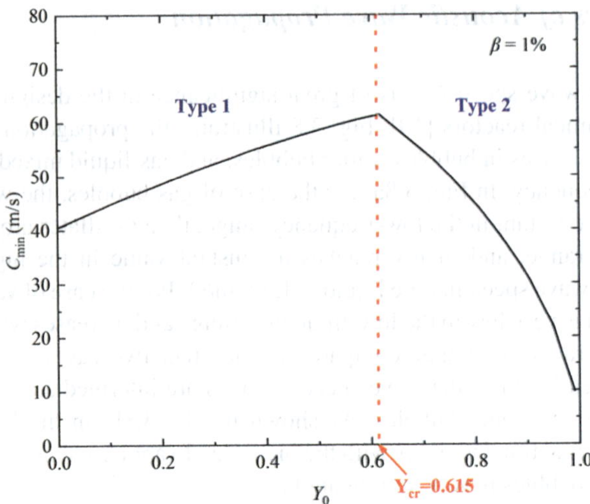

Fig. 3.9 The variation of the minimum wave speed c_{min} with the vapor mass fraction Y_0. $\beta = 1\%$. $R_0 = 100$ μm. (Reprinted with the permission from Ref. [19] Copyright (2018) (ELSEVIER))

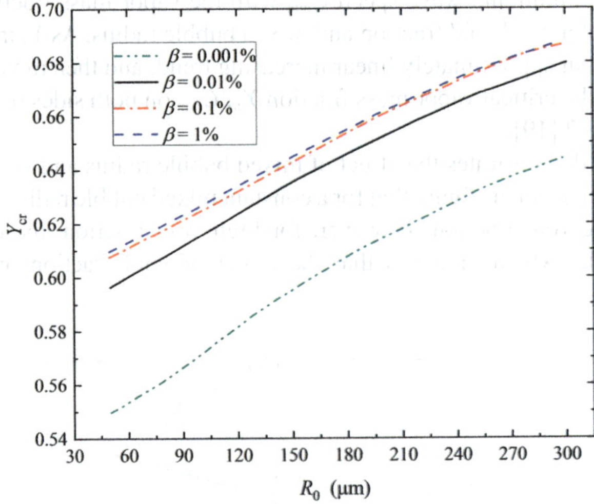

Fig. 3.10 Variation of critical vapor mass fraction Y_{cr} with mixed bubble radius and vapor mass fraction. $50 \leq R_0 \leq 300$ μm. (Reprinted with the permission from Ref. [19] Copyright (2018) (ELSEVIER))

limited. For a constant void fraction, Y_{cr} increases with the increase in the radius of the bubble. For the parameter regions studied, the values of Y_{cr} vary between 0.55 and 0.69 [19].

3.4 Critical Frequency

3.4.1 Low-frequency Critical Frequency

Figure 3.11 demonstrates the variation of the low critical frequency f_L with mixed bubble radius for different vapor mass fractions (void fraction $\beta = 0.10\%$). From Fig. 3.11, it can be seen that the vapor mass fraction Y_0 has a significant effect on f_L. When the mixed bubble radius is certain, f_L decreases with increasing Y_0. For small and medium vapor mass fractions (i.e., $Y_0 \leq 0.95$), f_L decreases monotonically with increasing mixed bubble radius. For large water vapor mass fractions (i.e., $Y_0 = 0.99$), when the mixed bubble radius is small ($R_0 \ll 100$ μm), f_L increases and then decreases with the presence of a local maximum. When the flagellum radius is large, f_L remains almost constant as the radius increases.

Figure 3.12 illustrates the variation of the low critical frequency f_L with void fraction for different vapor mass fractions (mixed bubble radius $R_0 = 100$ μm). At large void fraction ($\beta > 0.01\%$), f_L remains a constant for all Y_0, indicating that f_L is independent of void fraction. At small void fraction ($\beta < 0.01\%$), f_L decreases with increasing β for all Y_0, but the magnitude of the change is so small that it is almost negligible.

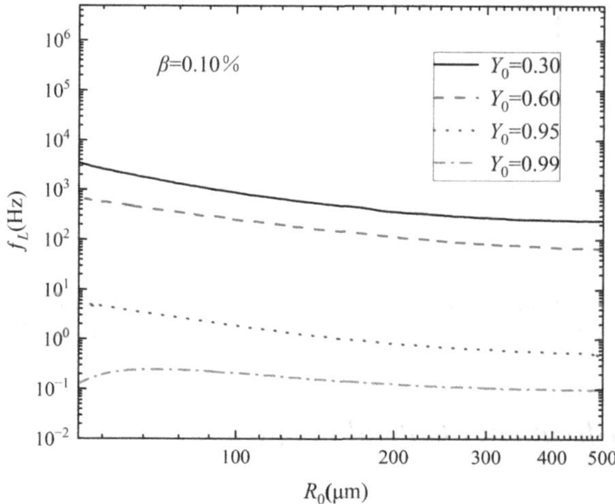

Fig. 3.11 The variation of the low critical frequency f_L with mixed bubble radius for different vapor mass fractions. $\beta = 0.10\%$. (Reprinted with the permission from Ref. [16] Copyright (2018) (ELSEVIER))

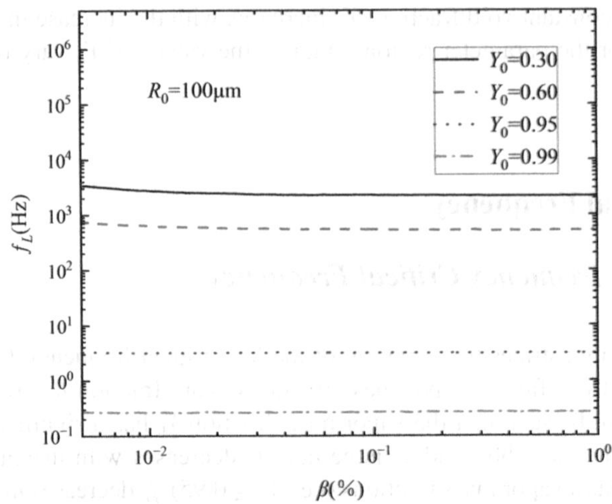

Fig. 3.12 Variation of low critical frequency f_L with void fraction for different vapor mass fractions. $R_0 = 100\,\mu\mathrm{m}$. (Reprinted with the permission from Ref. [16] Copyright (2018) (ELSEVIER))

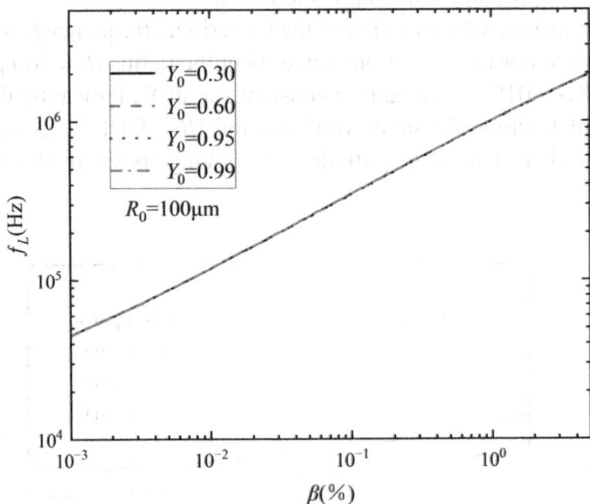

Fig. 3.13 The effect of void fraction and vapor mass fraction on the High-frequency critical frequency f_H. $R_0 = 100\,\mu\mathrm{m}$. (Reprinted with the permission from Ref. [16] Copyright (2018) (ELSEVIER))

3.4.2 High-frequency Critical Frequency

Figure 3.13 demonstrates the effect of void fraction and vapor mass fraction on the High-frequency critical frequency f_H. Figure 3.13 indicates that f_H exhibits a linear increase with β. However, the curves of f_H are identical for different Y_0, which indicates that f_H is independent of Y_0. This is consistent with the theoretical analysis.

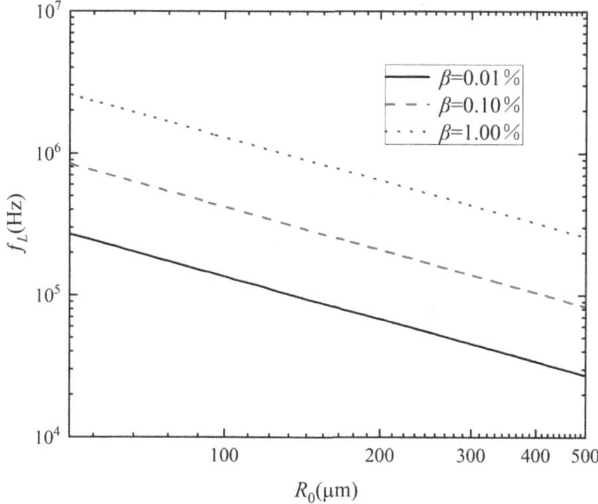

Fig. 3.14 The effect of void fraction f_H and mixed bubble radius R_0 on the High-frequency critical frequency f_H. (Reprinted with the permission from Ref. [16] Copyright (2018) (ELSEVIER))

Figure 3.14 demonstrates the effect of void fraction f_H and mixed bubble radius R_0 on the High-frequency critical frequency f_H. From the figure, it can be seen that as R_0 increases, f_H decreases linearly at all three void fractions. Keeping the radius constant, f_H increases with increasing void fraction for all.

3.5 Constant Wave Speed

3.5.1 Low-frequency Constant Wave Speed

Figures 3.15 and 3.16 display the variation of the low-frequency constant wave speed c_L with respect to the mixed bubble radius R_0 for different mass vapor fractions Y_0 at void fraction β of 0.10% and 1.00%, respectively. For $Y_0 \leq 0.95$, c_L remains a constant value. c_L is independent of the mixed bubble radius. For large Y_0, c_L increases gradually with increasing mixed bubble radius, and the increase is more pronounced for small mixed bubble radius ($R_0 \leq 100$ µm). This can be explained in conjunction with Eqs. (3.53) and (3.54), where surface tension has an effect only when Y_0 is large. When the radius of the mixed bubble is the same, the wave speed decreases with increasing vapor mass fraction Y_0. Figure 3.17 demonstrates the variation of the low-frequency constant wave velocity c_L with void fraction for the same vapor mass fraction. From the figure, it can be seen that c_L decreases linearly with increasing void fraction when $Y_0 \geq 0.30\%$. And when $Y_0 = 0.30\%$, at small void fraction β ($\beta \leq 0.001\%$), c_L decreases slowly with increasing void fraction, and subsequently, c_L decreases linearly with increasing void fraction.

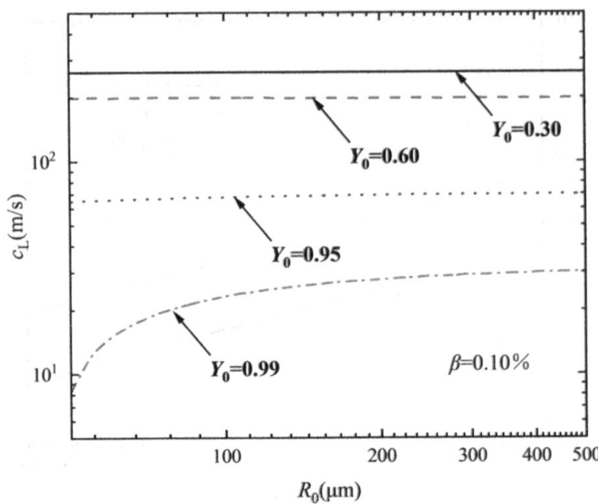

Fig. 3.15 Variation of low-frequency constant wave velocity c_L with mixed bubble radius for different vapor mass fractions. $\beta = 0.10\%$. (Reprinted with the permission from Ref. [16] Copyright (2018) (ELSEVIER))

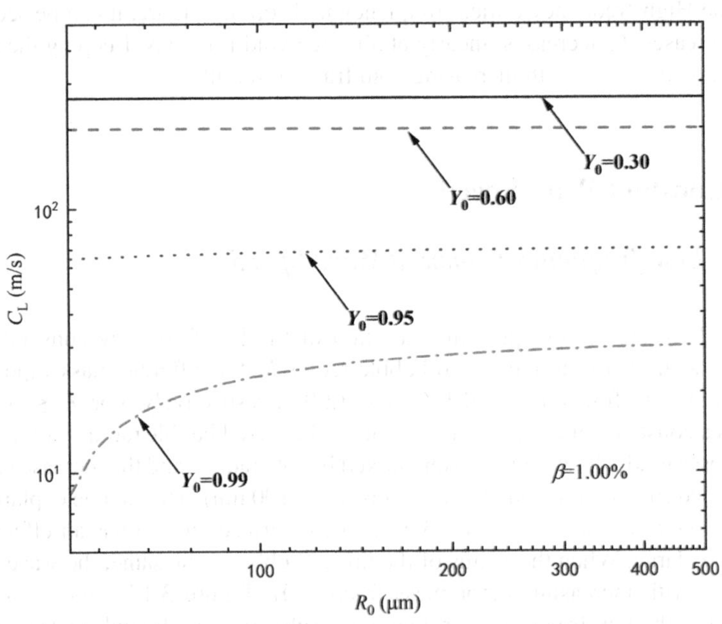

Fig. 3.16 Variation of low-frequency constant wave velocity c_L with mixed bubble radius for different vapor mass fractions. $\beta = 1.00\%$. (Reprinted with the permission from Ref. [16] Copyright (2018) (ELSEVIER))

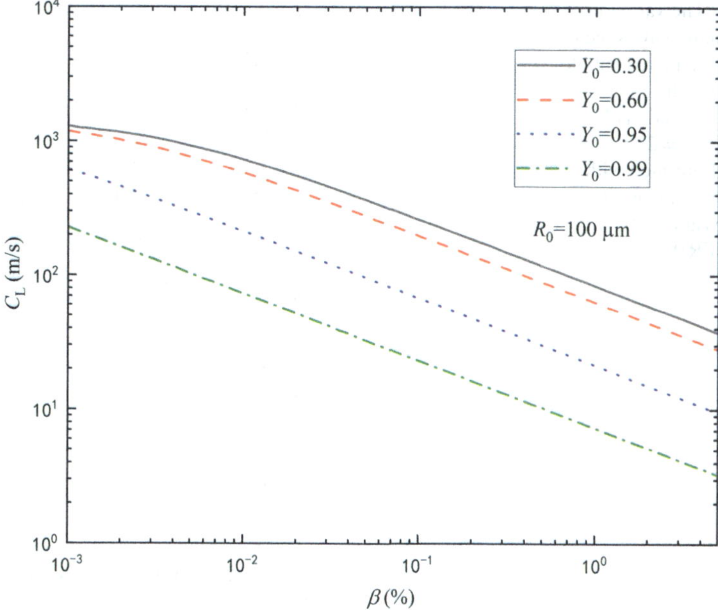

Fig. 3.17 Variation low-frequency constant wave velocity c_L with void fraction for the different vapor mass fraction Y_0. $R_0 = 100$ μm. (Reprinted with the permission from Ref. [16] Copyright (2018) (ELSEVIER))

3.5.2   High-frequency Constant Wave Speed

Combining Eq. (3.59) with Figs. 3.4, 3.5, 3.6 and 3.7, it can be seen that the high-frequency constant wave speed c_H is independent of the vapor mass fraction Y_0, the void fraction β and the mixed bubble radius R_0. Eq. (3.59) shows that the high-frequency constant wave speed c_H in a liquid is equal to the speed of acoustic wave in a pure liquid. In the derivation of the wave speed, water is assumed to be the surrounding liquid. Therefore, the constant high-frequency wave speed c_H is always equal to the speed of acoustic wave in water, i.e., $c_H = 1480$ m/s [16].

3.6   Minimum Wave Speed

Figure 3.18 demonstrates the variation of the minimum wave speed C_{min} with the mixed bubble radius R_0 for different vapor mass fractions Y_0. The void fraction β in sub-figures (a), (b) and (c) are 1%, 2% and 4%, respectively. From Fig. 3.18, it can be found that the variation of the minimum wave speed varies greatly at different vapor mass fractions, which indicates that the mass vapor fraction has a large effect on the minimum wave speed. By comparing the variation of minimum wave speed

Fig. 3.18 The variation of the minimum wave speed C_{min} with the mixed bubble radius R_0 for different vapor mass fractions Y_0. (**a**) $\beta = 1\%$; $\beta = 2\%$; $\beta = 4\%$. (Reprinted with the permission from Ref. [19] Copyright (2018) (ELSEVIER))

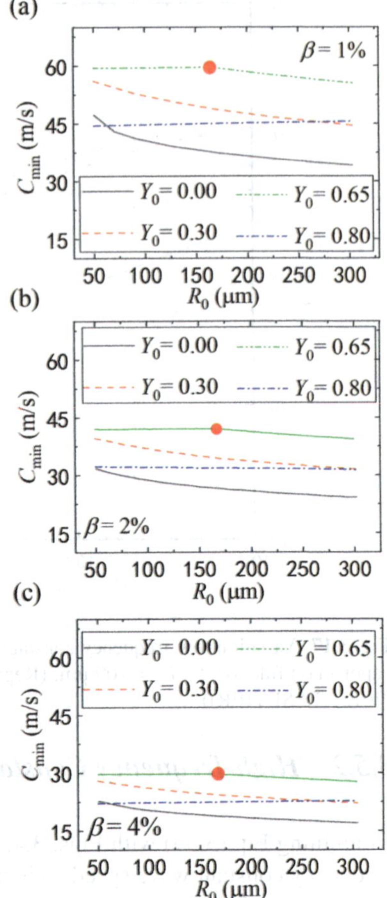

in the three sub-figures, it can be known that the minimum wave speed is decreasing with the increase of void fraction and the values of minimum wave speed tend to the same value.

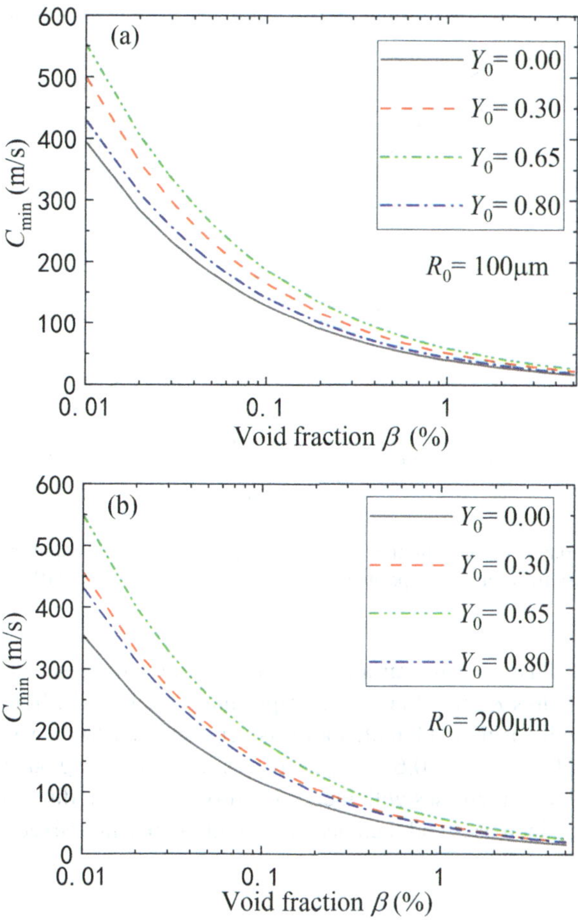

Fig. 3.19 The variation of the minimum wave speed C_{min} with void fraction for different vapor mass fractions Y_0. (**a**) $R_0 = 100\,\mu m$; (**a**) $R_0 = 200\,\mu m$. (Reprinted with the permission from Ref. [19] Copyright (2018) (ELSEVIER))

Figure 3.19 displays the variation of the minimum wave speed C_{min} with void fraction for different vapor mass fractions Y_0. The radii of the mixed bubbles in sub-figures (a) and (b) are 100 μm and 200 μm, respectively. It can be seen from the figures that C_{min}. decreases rapidly and then slowly as the void fraction increases and converges to almost the same value in the region of larger void fraction.

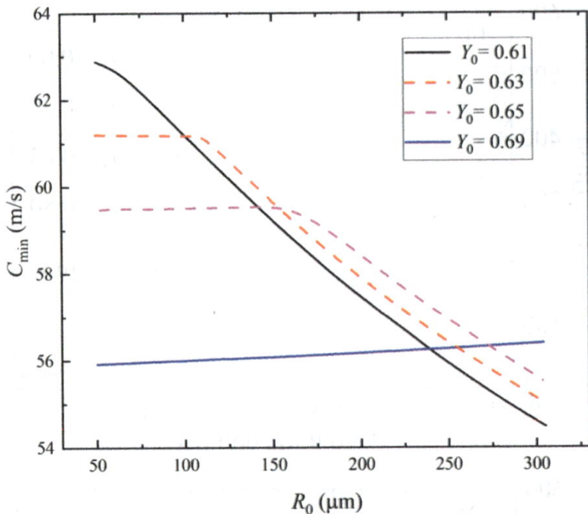

Fig. 3.20 The variation of the minimum wave speed C_{min} with the mixed bubble radius R_0 near the critical vapor mass fraction Y_{cr}. (Reprinted with the permission from Ref. [19] Copyright (2018) (ELSEVIER))

Figure 3.20 illustrates the variation of the minimum wave speed C_{min} with the mixed bubble radius R_0 near the critical vapor mass fraction Y_{cr}. In this region, the trend of C_{min} with gas mixed bubble radius is very sensitive to the vapor mass fraction. For example, when Y_0= 0.61, C_{min} decreases with increasing bubble radius, and when Y_0 = 0.69, C_{min} increases with increasing mixed bubble radius. For Y_0 between 0.61 ~ 0.69, C_{min} increases and then decreases with increasing mixed bubble radius.

References

1. Kargl SG. Effective medium approach to linear acoustics in bubbly liquids. J Acoust Soc Am. 2002;111(1):168–73.
2. Bader KB, Gruber MJ, Holland CK. Shaken and stirred: mechanisms of ultrasound-enhanced thrombolysis. Ultrasound Med Biol. 2015;41(1):187–96.
3. Weber TC. Observations of clustering inside oceanic bubble clouds and the effect on short-range acoustic propagation. J Acoust Soc Am. 2008;124(5):2783–92.
4. Li ZR, Sun L, Zong Z, et al. A boundary element method for the simulation of non-spherical bubbles and their interactions near a free surface. Acta Mech Sinica. 2012;28(1):51–65.
5. Nguyen QT, Nguyen VT, Phan TH, et al. Numerical study of dynamics of cavitation bubble collapse near oscillating walls. Phys Fluids. 2023;35(1):013306.
6. Wang X, Zhang C, Su H, et al. Research on cavitation bubble behaviors between a dual-particle pair. Phys Fluids. 2024;36(2):023310.
7. Wood A. A textbook of sound: being an account of the physics of vibrations with special reference to recent theoretical and technical developments. New York: The Macmillan company; 1941.
8. Brennen C. Cavitation and bubble dynamics. Oxford University Press; 1995.

9. Crespo A. Sound and shock waves in liquids containing bubbles. Phys Fluids. 1969;12(11):2274–82.
10. Ando K, Colonius T, Brennen CE. Improvement of acoustic theory of ultrasonic waves in dilute bubbly liquids. J Acoust Soc Am. 2009;126(3):EL69-EL74.
11. Prosperetti A. The speed of sound in a gas–vapour bubbly liquid. Interface focus. 2015;5(5):20150024.
12. Zhang Y-N, Guo Z-Y, Gao Y-H, et al. Valid regions of formulas of sound speed in bubbly liquids. Chin Phys Lett. 2017;34(6):064701.
13. Commander KW, Prosperetti A. Linear pressure waves in bubbly liquids: comparison between theory and experiments. J Acoust Soc Am. 1989;85(2):732–46.
14. Zhang Y. Heat transfer across interfaces of oscillating gas bubbles in liquids under acoustic excitation. Int Commun Heat Mass Transfer. 2013;43:1–7.
15. Zhang Y, Li S. The secondary Bjerknes force between two gas bubbles under dual-frequency acoustic excitation. Ultrason Sonochem. 2016;29:129–45.
16. Zhang Y, Guo Z, Du X. Wave propagation in liquids with oscillating vapor-gas bubbles. Appl Therm Eng. 2018;133:483–92.
17. Zhang Y, Gao Y, Guo Z, et al. Effects of mass transfer on damping mechanisms of vapor bubbles oscillating in liquids. Ultrason Sonochem. 2018;40:120–7.
18. Fuster D, Montel F. Mass transfer effects on linear wave propagation in diluted bubbly liquids. J Fluid Mech. 2015;779:598–621.
19. Zhang Y, Guo Z, Gao Y, et al. Acoustic wave propagation in bubbly flow with gas, vapor or their mixtures. Ultrason Sonochem. 2018;40:40–5.

Chapter 4
Cavitation in Vapor/Liquid/Solid Multiphase Flow

This chapter examines the phenomena associated with cavitation bubble collapse, jet, and shock wave propagation in liquids containing sand and cavitation bubbles. First the collapse behavior of cavitation bubble near particles is described, followed by an analysis of the jet generation mechanism. Then, the generation and the propagation of shock waves during the cavitation bubble oscillation near particles are discussed. Finally, the cavitation and erosion of hydraulic machinery under the synergistic effect of particles and cavitation bubbles is discussed.

4.1 Bubble-Water-Particle System

4.1.1 Experimental System

Figure 4.1 illustrates the experimental schematic of a laser-induced cavitation bubble near particles. The cavitation vesicle is generated in deionized water after the laser beam generated by the Nd: YAG laser generator passes through the focusing lens. Due to the inconvenient movement of the laser generator, the location of the cavitation bubble generation is determined by the focusing lens. During the experiment, firstly, the incipient position of the cavitation bubble and its distance from the particles should be determined. Specifically, the position of the cavitation bubble is determined by the preview system of the high speed camera, and the particles are moved to the designed position by the 3-D platform. The particle is fixed on a fine needle using adhesive. The needle is positioned on the side away from the cavitation bubble inception, rendering its influence on the bubble negligible. Then, the recording function of the high-speed camera is turned on and the laser generator is controlled to emit a laser beam by triggering the digital delay generator. The light source is used to increase the brightness of the image, and a computer stores the morphological changes of the cavitation bubble recorded by the high-speed camera.

© The Author(s), under exclusive license to Springer Nature Switzerland AG 2025

J. Hu et al., *Multiphase Flow with Bubbles*, SpringerBriefs in Energy, https://doi.org/10.1007/978-3-031-99216-2_4

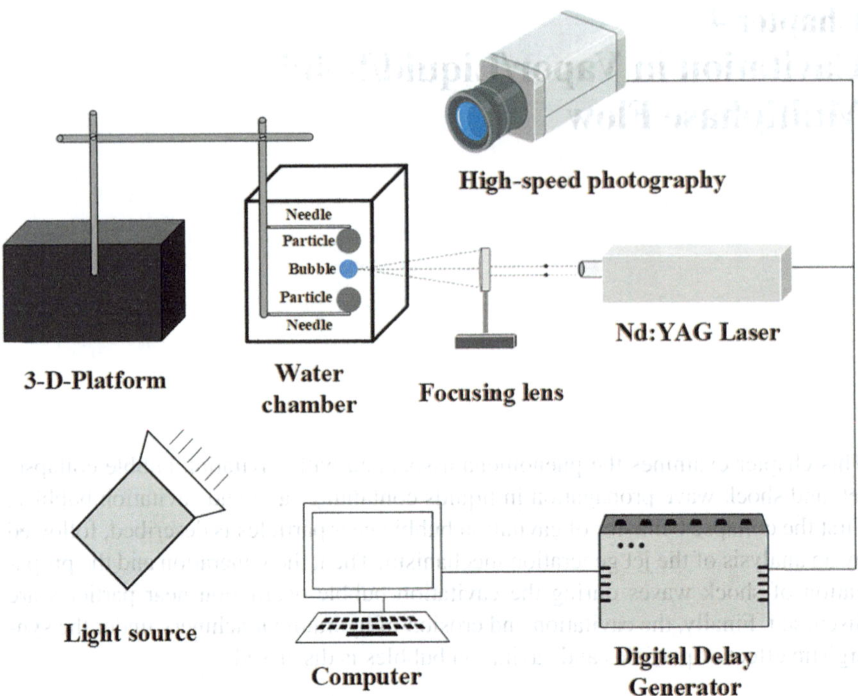

Fig. 4.1 Experimental schematic of a laser-induced cavitation bubble near particles. (Reprinted with the permission from Ref. [1] Copyright (2025) (ELSEVIER))

4.1.2   Single Particle

Three cases are identified based on the morphology of cavitation bubble collapse, which are mushroom-shaped (case 1), pear-shaped (case 2), and sphere-shaped (case 3). Figure 4.2 illustrates the collapse behavior of a laser-induced cavitation bubble near a SiO_2 spherical particle for case 1. After the laser beam passes through the focusing lens, a high-energy laser spot is foamed near the particle (as seen in frame 2). The laser ionizes the deionized water, producing a cavitation bubble that contains a large amount of vapor along with a small quantity of non-condensable gas. Due to the pressure and the temperature inside the cavitation bubble being significantly higher than those in the surrounding water, the bubble begins to grow. From frames 2 to 9, the cavitation bubble expands under inertial forces. Since the bubble inception at a short distance from the particle, its left side develops along the particle surface, while the right side, being less influenced by the particle, grows spherical. When the cavitation bubble reaches its maximum volume (frame 9), its interfacial curvature changes significantly, with the greatest curvature observed at the position indicated by a red dashed circle.

From frames 11 to 15, the internal pressure of the cavitation bubble falls below that of the surrounding water, and the bubble begins to collapse. According to the

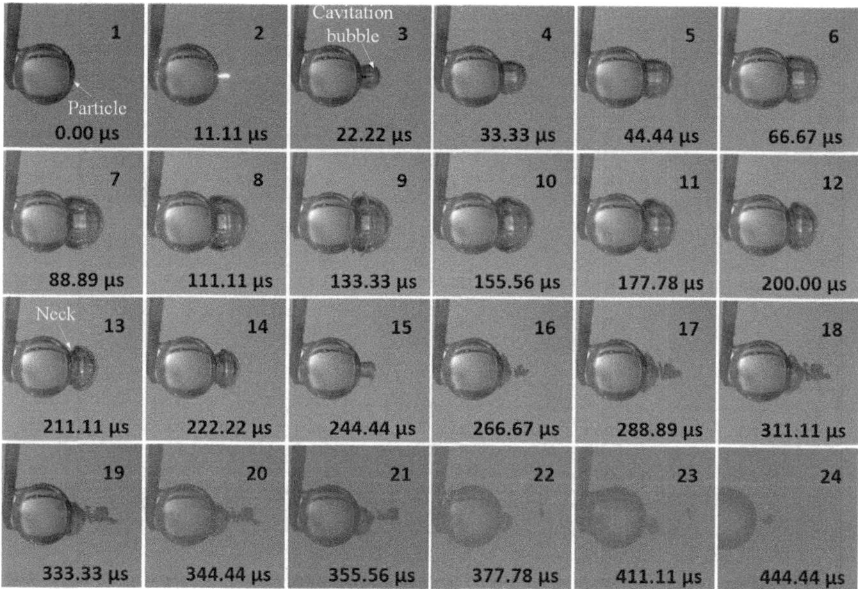

Fig. 4.2 Collapse behavior of a laser-induced cavitation bubble near spherical particle for case 1. The red dashed circle indicates the location of the bubble with the greatest curvature. (Reprinted with the permission from Ref. [3] Copyright (2018) (ELSEVIER))

research by Koch et al. [2], regions of the cavitation bubble with higher curvature contract more rapidly due to flow focusing. Consequently, a neck forms on the left interface of the bubble, while the right side collapses in a spherical shape. In frame 15, the bubble is observed contracting to its minimum volume for the first time, with its internal pressure and temperature once again exceeding those of the surrounding water, and it is about to rebound.

During the rebound stage depicted in frames 16 to 19, the cavitation bubble splits into two parts. The portion adjacent to the particle rebounds along the surface of the particle, whereas the other part gradually moves away from it. From frames 20 to 24, the cavitation bubble re-collapses and gradually vanishes. A comparison between the state of the bubble in frames 2 to 15 (the first period) and frames 16 to 24 (the second period) reveals that both the maximum volume and the duration in the first period are greater than those in the second period. This difference arises because, during the first period, most of the bubble energy (both pressure and heat) is transferred to the surrounding water, thereby reducing the energy available for the second period.

Figure 4.3 presents the collapse behavior of a cavitation bubble for case 2. During the early growth stage of the cavitation bubble (frames 1–4), the bubble expands nearly spherically. During the subsequent growth stage (frames 5–8), the left side of the cavitation bubble is influenced by the particle and gradually conforms to its surface. During the collapse stage (frames 9–15), the left side of the cavitation bubble contracts along the particle's surface with minimal shrinkage, while the right

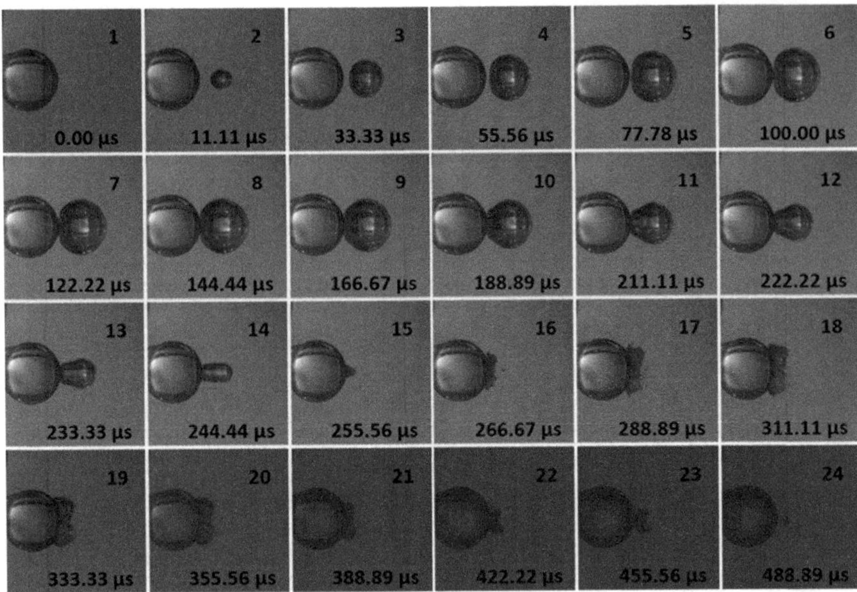

Fig. 4.3 Collapse behavior of a laser-induced cavitation bubble near spherical particle for case 2. (Reprinted with the permission from Ref. [3] Copyright (2018) (ELSEVIER))

side, uninfluenced by any physical boundary, rapidly collapses in a spherical shape. Throughout this process, the shape of the cavitation bubble transitions from a pear shape (frames 10–13) to a cylindrical shape (frame 14), and it first reaches its minimum volume at the right endpoint of the particle (frame 15). During the rebound stage (frames 16–19), the rapid collapse of the bubble's right side during the collapse stage suppresses rebound in the axial direction (the line connecting the centers of the particle and the cavitation bubble), resulting in a more pronounced rebound in other directions. In frames 20–24, the cavitation bubble collapses once again. Compared to Fig. 4.2, the particle causes the cavitation bubble to collapse in its vicinity without inducing bubble splitting. Moreover, the surface smoothness of the cavitation bubble in the second period is significantly reduced compared to that in the first period.

Figure 4.4 illustrates the collapse behavior of a cavitation bubble at a long distance from the particle for case 3. During the first period (frames 1–13), the bubble expands and collapses in an almost spherical shape while its center gradually moves close to the particle. During the second (frames 14–20) and third periods (frames 21–24), the bubble exhibits a decrease in sphericity and surface smoothness, and it continues moving further toward the particle.

Figure 4.5 shows the relationship between the collapse behavior of a cavitation bubble and two key parameters. These key parameters are defined as the maximum radius of the cavitation bubble, R_{max}, and the distance, d, between the inception position of the cavitation bubble and the center of the particle. The influence of the

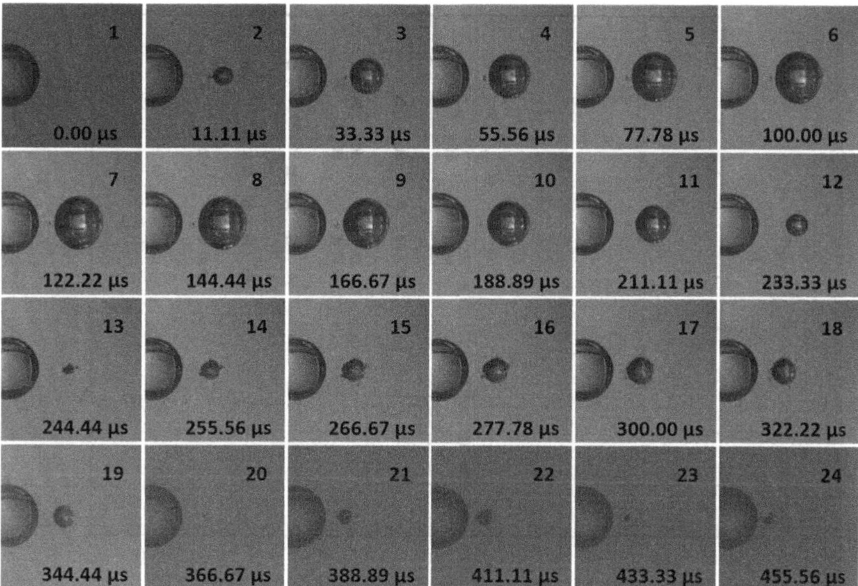

Fig. 4.4 Collapse behavior of a laser-induced cavitation bubble near spherical particle for case 3. (Reprinted with the permission from Ref. [3] Copyright (2018) (ELSEVIER))

distance parameter is greater than that of R_{max}. In the figure, χ is defined as the ratio of R_{max} to d, with χ_1 representing the critical condition distinguishing case 1 from case 2 and χ_2 serving as the criterion for differentiating case 2 from case 3. For particle radii of 0.5, 1.0, and 1.5 mm, χ_1 is 1.00, 0.64, and 0.54 respectively, while χ_2 is 0.82, 0.45, and 0.39 respectively. Both χ_1 and χ_2 decrease with increasing particle radius.

4.1.3 Double Particles

Three cases are identified based on the morphology of cavitation bubble collapse, which are gyroscopic shape, drum shape, olive shape and spherical shape. Figure 4.6 illustrates the collapse of a cavitation bubble between two spherical particles in a gyroscopic shape. When the bubble is at its maximum size (frame 1), its upper and lower portions conform to the surfaces of the two particles, respectively. Due to flow focusing, the bubble section at the particle interface (outlined by a red dotted circle) contracts the fastest and forms two necks. Consequently, from frames 2 to 10, the two necks gradually move toward each other, causing the bubble to collapse in a gyroscopic shape. In frame 11, after the collision of the two necks, an annular depression forms in middle of the bubble, which nearly contracts horizontally

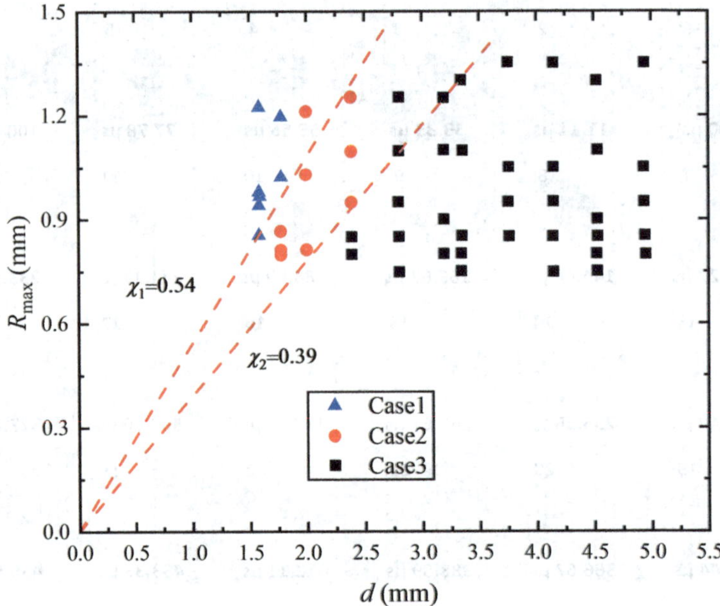

Fig. 4.5 Relationship between the collapse behavior of a cavitation bubble and the key parameters R_{max}, and d. (Reprinted with the permission from Ref. [3] Copyright (2018) (ELSEVIER))

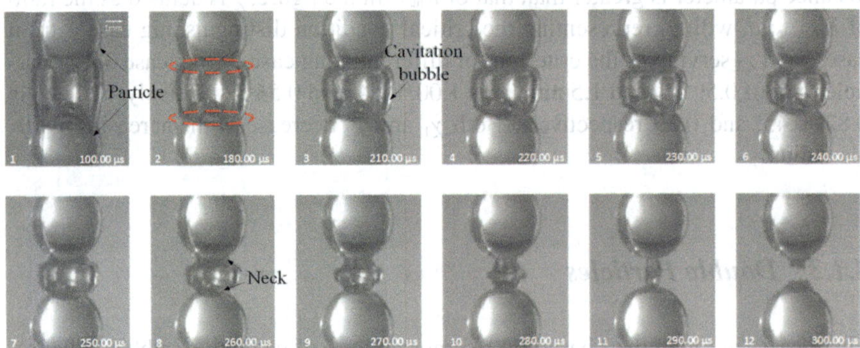

Fig. 4.6 The cavitation bubble collapses in a gyroscopic shape between two spherical particles. (Reprinted with the permission from Ref. [4] Copyright (2023) (Springer))

toward the center and splits the bubble into two parts. In frame 12, the two split cavitation bubbles collapse separately near each of the particles.

Figure 4.7 illustrates the collapse of a cavitation bubble between two spherical particles in a drum shape. When the bubble reaches its maximum size (frame 1), its upper and lower portions are in contact with the two particles, although the contact area is smaller than that shown in Fig. 4.6. From frames 2 to 10, the sections of the bubble near the particles contract slowly while the central portion contracts rapidly,

Fig. 4.7 The cavitation bubble collapses in a drum shape between two spherical particles. (Reprinted with the permission from Ref. [4] Copyright (2023) (Springer))

Fig. 4.8 The cavitation bubble collapses in an olive shape between two spherical particles. (Reprinted with the permission from Ref. [4] Copyright (2023) (Springer))

resulting in a drum shape collapse. In frame 11, the contraction rate in the bubble's central region accelerates, producing an annular depression. The formation mechanism of this annular depression is distinct from that observed in frame 11 of Fig. 4.6. The contraction of the annular depression leads to the splitting of the bubble, and the two resulting cavitation bubbles collapse near the particles to which they are closest.

Figure 4.8 illustrates the collapse of a cavitation bubble between two spherical particles in an olive shape. Since the spacing between the two particles is slightly larger than the maximum diameter of the cavitation bubble, the bubble does not contact either particle at its maximum size. Under the influence of the particles, there is a significant difference in the contraction velocity along the bubble interface. Specifically, the contraction velocity at the top and the bottom ends are slow, while the contraction velocity in the middle is fast. From frames 2 to 10, it is observed that the contraction distance perpendicular to the bubble center is small at the top and the bottom ends and large at other positions, resulting in an olive shape

Fig. 4.9 The cavitation bubble collapses in a spherical shape between two spherical particles. (Reprinted with the permission from Ref. [4] Copyright (2023) (Springer))

during collapse. In frame 11, slight depressions appear at the upper and the lower ends, and in frame 12, the bubble reaches its minimum size for the first time.

Figure 4.9 demonstrates the collapse of a cavitation bubble between two spherical particles in a spherical shape. Since the distance between the particles is much greater than the maximum diameter of the cavitation bubble, the particles exert minimal influence on the bubble. The pressure gradient in the surrounding fluid is nearly uniform, resulting in an almost perfect spherical collapse of the cavitation bubble.

Figure 4.10 illustrates the relationship between the collapse behavior of a cavitation bubble between two particles and the key parameters R_{max} and d. Here, R_{max} is defined as the maximum radius of the cavitation bubble, while d represents the minimum distance between the bubble's inception position and the particle surface. It is evident that the distance d exerts the most significant influence on the bubble collapse behavior. As this distance d increases, the compressive effect of the particles diminishes, leading to a sequential transition of the bubble collapse shape from gyroscopic to drum, then to olive, and finally to spherical. The influence of R_{max} on the collapse behavior is relatively minor. The parameter ξ, defined as the ratio of d to R_{max}, characterizes the critical conditions for these shape transitions. The critical conditions for the collapse from gyroscopic to drum shape, from drum to olive shape, and from olive to spherical collapse are $\xi = 0.27$, 1.17, and 1.56, respectively.

4.1.4   Triple Particles

This section describes the collapse behavior of a cavitation bubble incipient along at the altitude line when three spherical particles are positioned at the vertices of an equilateral triangle. Three typical scenarios have been identified based on the collapse behavior of the cavitation bubble and are exemplified in Figs. 4.11, 4.12, and

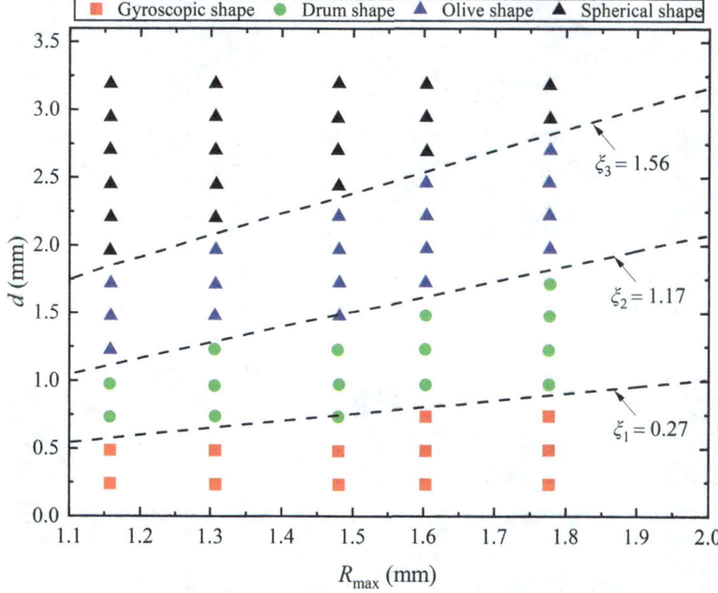

Fig. 4.10 Relationship between the collapse behavior of a cavitation bubble between two particles and the key parameters R_{max} and d. (Reprinted with the permission from Ref. [4] Copyright (2023) (Springer))

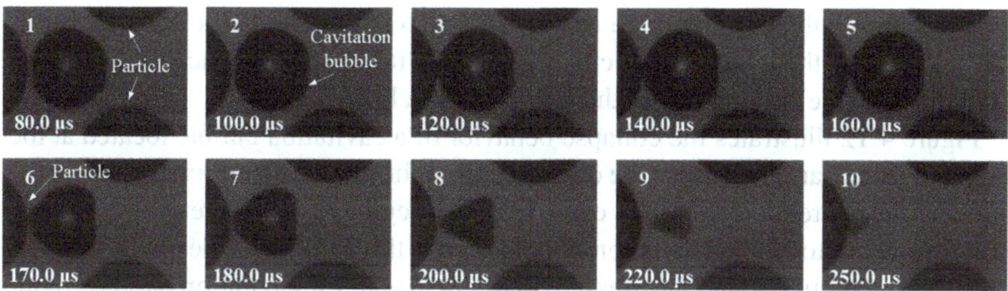

Fig. 4.11 Collapse behavior of the cavitation bubble at a short distance to the leftmost particle. (Reprinted with the permission from Ref. [5] Copyright (2024) (AIP Publishing))

4.13. From Figs. 4.11, 4.12 and 4.13, the maximum radius of the cavitation bubble and the distance between particles remain constant, but the distance between the bubble inception and the leftmost particle increases.

Figure 4.11 illustrates the collapse behavior of the cavitation bubble at a short distance to the leftmost particle. The leftmost particle exerts the greatest influence on the bubble, while the other two particles have a comparatively lesser effect. In frames 1–5, the portion of the bubble near the leftmost particle collapses the slowest, with small differences in the rate of collapse at the remaining positions. From frames 6 to 9, the two particles on the right slow down the collapse of the bubble in their vicinity, and the bubble transforms into a conical shape. Since the right side of

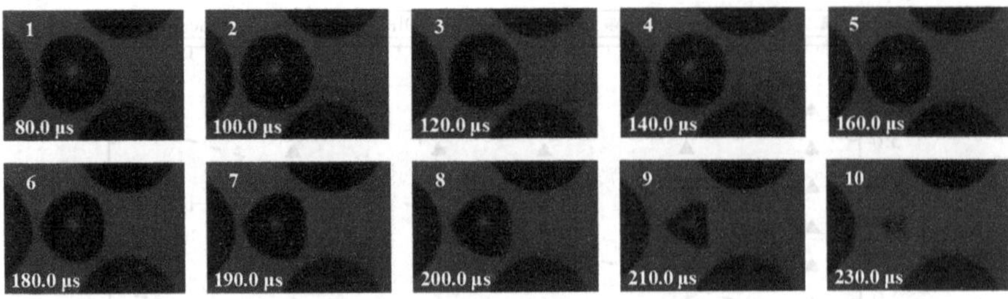

Fig. 4.12 Collapse behavior of a cavitation bubble located at the center of three particles. (Reprinted with the permission from Ref. [5] Copyright (2024) (AIP Publishing))

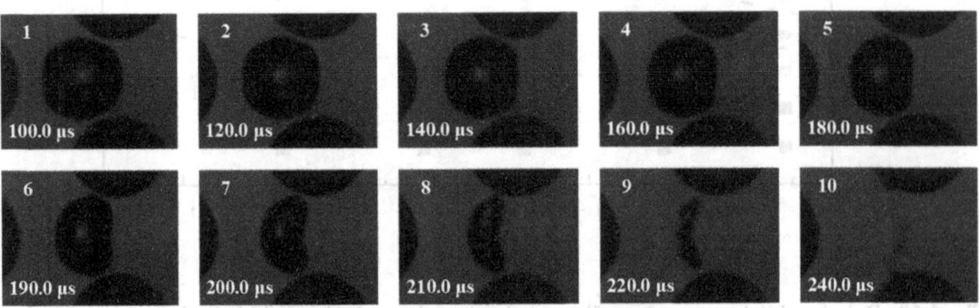

Fig. 4.13 Collapse behavior of the cavitation bubble at a long distance from the leftmost particle. (Reprinted with the permission from Ref. [5] Copyright (2024) (AIP Publishing))

the cavitation bubble on the high line is least affected by the particles, a jet forms directed toward the leftmost particle. In frame 10, under the influence of the jet, the cavitation bubble collapses near the leftmost particle.

Figure 4.12 illustrates the collapse behavior of a cavitation bubble located at the center of three particles. Since the distances from the bubble inception to each of the three particles are identical, they exert an equal degree of influence on the bubble. From frames 1 to 8, the cavitation bubble on the three altitude lines is little influenced by the particles, and the pressure gradient toward the cavitation bubble gradually increases. Consequently, as shown in frame 9, the cavitation bubble on the altitude line produces a depression toward the center. In frame 10, the cavitation bubble collapses near the center of the three particles.

Figure 4.13 illustrates the collapse behavior of the cavitation bubble at a long distance from the leftmost particle. The collapse of the bubble near the two right particles is most inhibited and they show minimal displacement toward the bubble center during collapse. The leftmost particle has little effect on the cavitation bubble, and the left side of the cavitation bubble collapses in a nearly spherical shape. In frames 6 to 8, the right side of the cavitation bubble on the altitude line is least affected by the particles and gradually develops a jet toward the leftmost particle. In frame 9, the jet pierces the left interface of the cavitation bubble. In frame 10, the cavitation bubble collapses completely.

4.2 Jet Near Particles

This section describes the jet phenomenon generated by the cavitation bubble collapse near particles. The mechanism of jet formation and evolution is revealed by the pressure and the velocity distributions on the slice passing through the axis of the particle and the cavitation bubble.

4.2.1 Single Jet

Fig. 4.14 shows a single jet produced by the collapse of a cavitation bubble near a single spherical particle. In frames 1–2, a uniformly distributed pressure gradient forms around the cavitation bubble. Under the influence of the pressure gradient directed toward the cavitation bubble, it begins to collapse. In frames 3–4, the particle causes the pressure gradient distribution to become uneven. In particular, the pressure gradient on the right side of the cavitation bubble gradually increases and aggregates toward its right end. Subjected to the effect of the aggregated pressure gradient, a jet directed toward the particle forms on the right side of the cavitation bubble. In frame 7, the jet impacts on the right end of the particle results in a sudden increase in pressure. As the jet continues to impact, as shown in frame 8, the impact-induced pressure decreases.

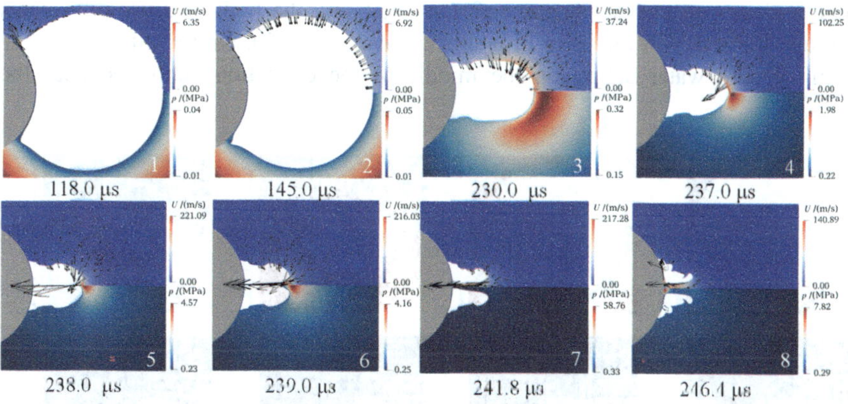

Fig. 4.14 A single jet produced by the collapse of a cavitation bubble near a single particle. The gray and white parts represent particles and cavitation bubbles, respectively. The upper part of each subplot is the velocity distribution and the lower one is the pressure distribution. (Reprinted with the permission from Ref. [6] Open access under a CC BY 4.0 license, https://creativecommons. org/licenses/by/4.0/)

4.2.2 Multiple Jets

Figure 4.15 shows the formation of dual jets produced by the collapse of a cavitation bubble near a single particle. In frame 1, after the low-pressure fluid in the gap between the particle and the cavitation bubble flows out, it moves toward the part of the cavitation bubble that is slightly farther from the particle, while the fluid on the right side of the bubble is already flowing toward the bubble. The presence of the particle causes the contraction on the left side of the cavitation bubble to lag that on the right side. In frames 2–3, the pressure gradient of the fluid surrounding the cavitation bubble gradually increases, causing the bubble to begin contracting toward the center. In frames 4–5, the high-pressure fluid at the depression on the left side of the cavitation bubble promotes the formation of a jet directed away from the particle. On the right side, the continuously converging and intensifying high-pressure fluid along the symmetric axis leads to the formation of a jet directed toward the particle. In frames 6–7, due to the difference in the intensity of the pressure gradients, the jet on the right is stronger than the one on the left. In frame 8, after the two jets collide, the fluid within them continues to move toward the particle.

Figure 4.16 shows the bubble splitting behavior and multi-jet phenomena of a cavitation bubble between two particles. From Figs. 4.16a–d, the distance between the particles and the cavitation bubble increases. In Fig. 4.16a, the portion of the cavitation bubble adjacent to the particle first contracts and develops a neck structure (frames 1–2). The two necks then collide in the middle of the cavitation bubble, creating a high-pressure region at the collision site. The water in high-pressure region gives rise to two phenomena. On one hand, the high pressure propagates through the water in the form of a shock wave (frame 3). On the other hand, the high-pressure water acting on the middle of the cavitation bubble induces the

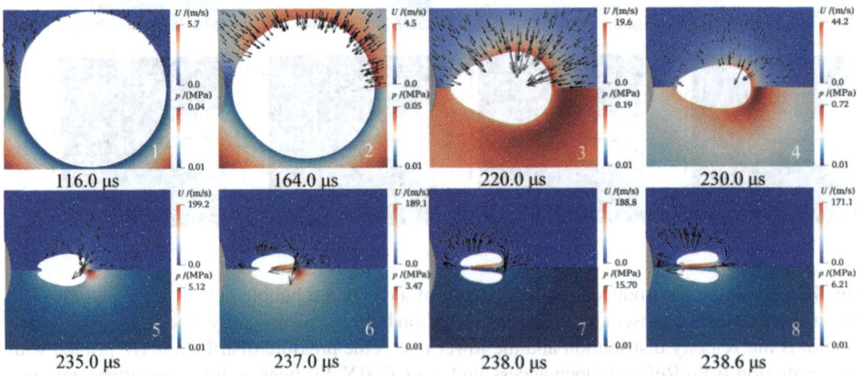

Fig. 4.15 The formation of dual jets produced by the collapse of a cavitation bubble near a single particle. The gray and white parts represent particles and cavitation bubbles, respectively. The upper part of each subplot is the velocity distribution and the lower one is the pressure distribution. (Reprinted with the permission from Ref. [6] Open access under a CC BY 4.0 license, https://creativecommons.org/licenses/by/4.0/)

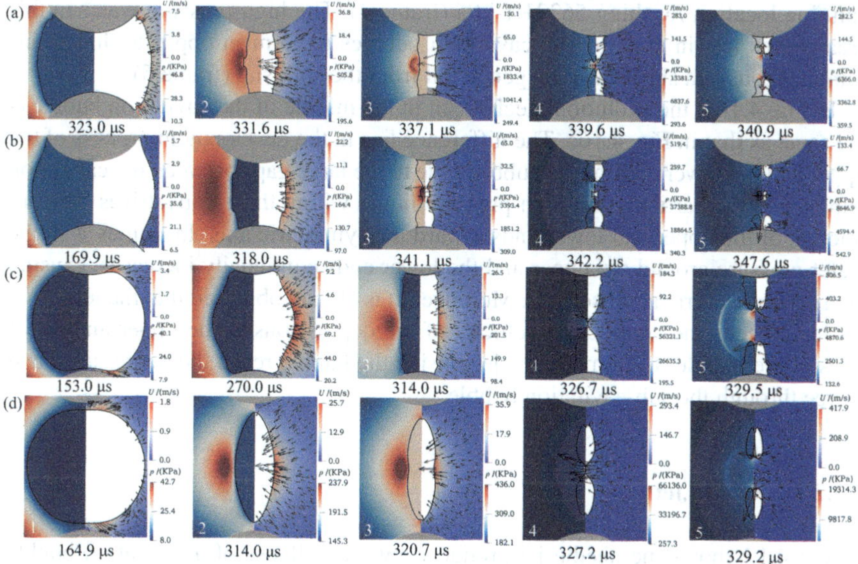

Fig. 4.16 The bubble splitting behavior and multi-jet phenomena of a cavitation bubble between two particles. The gray and white parts represent particles and cavitation bubbles, respectively. The left part of each subplot is the pressure distribution and the right one is the velocity distribution. (Reprinted with the permission from Ref. [7] Copyright (2023) (ELSEVIER))

formation of a rapidly contracting annular depression. When the annular depression converges onto the symmetric axis, the cavitation bubble splits into two sub-cavitation bubbles, with a high-pressure region generated at the split (frame 4). This high-pressure region drives the contraction of the bubble to form two jets while pushing the two sub-cavitation bubbles toward the particles.

In Fig. 4.16b, due to the slightly increased distance between the particles and the cavitation bubble, the two necks do not collide but instead each contract toward the symmetric axis (frames 1–3). Subsequently, the cavitation bubble splits into three parts, with high-pressure regions forming at the splitting points (frame 4). The water from the two high-pressure regions produces jets directed toward the particles within the upper and lower sub-cavitation bubbles. Meanwhile, under the combined influence of the water in the two high-pressure regions, a pair of oppositely directed jets forms at the upper and lower ends of the central part of the cavitation bubble.

In Fig. 4.16c, with an increased distance between the cavitation bubble and the particles, no neck structure forms during the bubble collapse. In frames 1–2, since the water pressure surrounding the middle of the cavitation bubble is higher than that at other regions, the bubble middle contracts rapidly. In frame 3, a relatively concentrated annular high-pressure water region appears near the bubble middle. Under the influence of this annular high-pressure region, an annular depression develops at the center of the cavitation bubble. When this annular depression contracts onto the symmetric axis, the cavitation bubble splits, with the pressure at the splitting location reaching 56321.1 kPa (frame 4). The high-pressure water in this area causes jets in the two sub-cavitation bubbles to form in opposite directions. Simultaneously, a shock wave is generated from the split site (frame 5).

In Fig. 4.16d, the evolution mechanism in the middle of the cavitation bubble is the same as that in Fig. 4.16c. The difference lies in the development process of the upper and the lower ends of the bubble. Due to the large gaps between the cavitation bubble and the particles, the high-pressure liquid within these gaps drives the contraction of the upper and lower ends toward the symmetric axis. In frames 3–4, the water pressure around the upper and the lower ends continually increases, causing depressions to form at those ends. Moreover, it is clearly observed that the jet velocity far exceeds the contraction velocity of the depressions at the upper and lower ends of the bubble. In frame 5, the jets inside the bubble reach the depressions and pierce through the sub-cavitation bubble.

4.2.2.1   Needle Jet

Figure 4.17 shows the needle jet produced by the collapse of a cavitation bubble between a particle and a wall. In Fig. 4.17a, the particle is close to the wall and the cavitation bubble is squeezed by both. The upper part of the cavitation bubble develops into a hemispherical shape along the particle surface, while the lower part becomes flattened along the wall. Since the particle is smaller in scale than the wall, its influence on the cavitation bubble is less significant compared to that of the wall. The cavitation bubble near the particle surface contracts the fastest, whereas the contraction is slowest near the wall. In frame 3, the cavitation bubble collapses at the lower end of the particle, producing a small-sized, high-pressure region in the water. Since the high-pressure region is much smaller than the cavitation bubble, a jet with a small diameter and high velocity forms in the upper part of the bubble; the

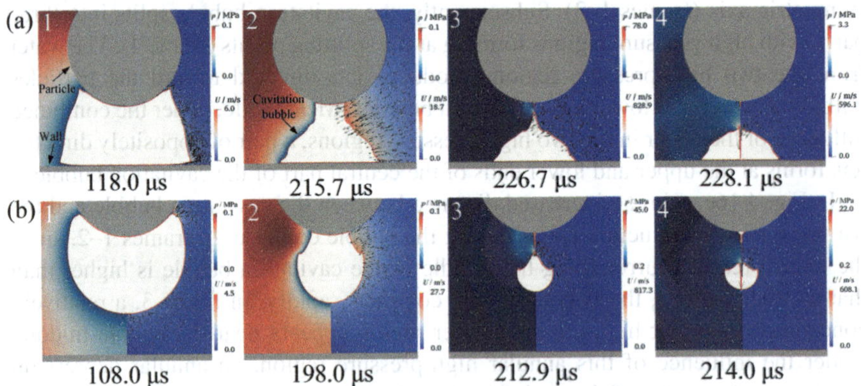

Fig. 4.17 Needle jet produced by the collapse of a cavitation bubble between a particle and a wall. (Reprinted with the permission from Ref. [8] Copyright (2023) (AIP Publishing))

jet is referred to as a needle jet. In frame 4, the needle jet impacts the wall at an extremely high velocity.

In Fig. 4.17b, the cavitation bubble is close to the particle but far from the wall. The cavitation bubble first contracts along the particle surface to form a neck (frame 2). When the bubble contracts to the lower end of the particle, a collision occurs (frame 3). The high-pressure water generated at the collision zone acts on the cavitation bubble, leading to the formation of a needle jet. In frame 4, the needle jet pierces through the lower end of the cavitation bubble and begins to develop toward the wall.

4.3   Shock Waves Near Particles

4.3.1   Nucleation

Figure 4.18 illustrates the evolution of the shock wave generated by a laser-induced cavitation bubble. Using a high-velocity double-exposure imaging technique, nanosecond time-resolved images were acquired to capture the formation and the propagation of the shock wave. After the laser pulse triggers optical breakdown in water, the rapidly formed plasma excites the cavitation bubble and immediately generates a high-velocity shock wave. From Figs. 4.18a–f, it can be observed that, since the propagation velocity of the shock wave front far exceeds the velocity of the material within the cavitation bubble, the shock wave quickly detaches from the periphery of both the plasma and the cavitation bubble. The wavefront of the shock wave exhibits

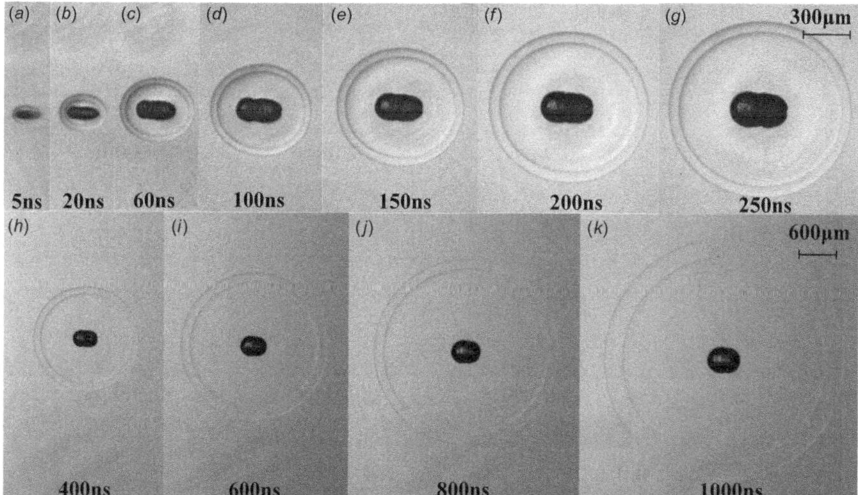

Fig. 4.18 Evolution of the shock wave generated by a laser-induced cavitation bubble. (Reprinted with the permission from Ref. [9] Copyright (2021) (ASME))

a markedly asymmetric (ellipsoidal) structure. This nonspherical characteristic may be contributed to factors such as the laser focusing angle, which results in an inhomogeneous plasma shape and energy distribution, causing different propagation velocity of the shock wave along various directions. In Figs. 4.18g–k, due to the energy dissipation of the shock wave and the influence of ambient pressure, the evolution of shock wave stabilizes and propagates in a spherical shape.

Figure 4.19 demonstrates the propagation of the shock wave at the inception of a cavitation bubble near a single particle. The numerical schlieren method is employed to visualize the shock wave. After the cavitation bubble inception, the high-pressure inside propagates outward in the form of a wave. When the wave impacts the particle, it is reflected, forming a reflected shock wave that travels toward the cavitation bubble (frame 4). Due to the significant difference in the fluid densities on either side of the cavitation bubble interface, the reflected wave is again reflected upon reaching the interface (frames 5–6). In frames 7–8, multiple reflections of the shock wave between the cavitation bubble interface and the particle can be observed.

Figure 4.20 illustrates the effect of particle to cavitation bubble distance on the variation of particle apex pressure induced by the shock wave. The dimensionless parameter p^* is defined as the ratio of the particle apex pressure to the ambient pressure (101,325 Pa). The shock wave causes a significant increase in the pressure on the particle surface, and the amplitude of the increase due to energy dissipation decreases with increasing dimensionless distance ξ. The shock wave impact causes the particle surface pressure to first rise and then falls, with the duration of the pressure rise being shorter than that of the pressure drop. As the dimensionless distance

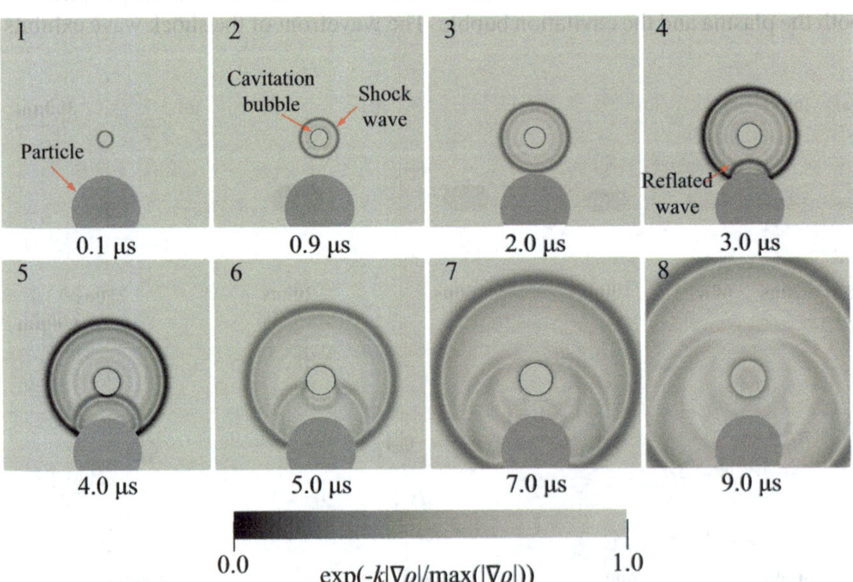

Fig. 4.19 Propagation of the shock wave at the inception of a cavitation bubble near a single particle. (Reprinted with the permission from Ref. [10] Copyright (2024) (ELSEVIER))

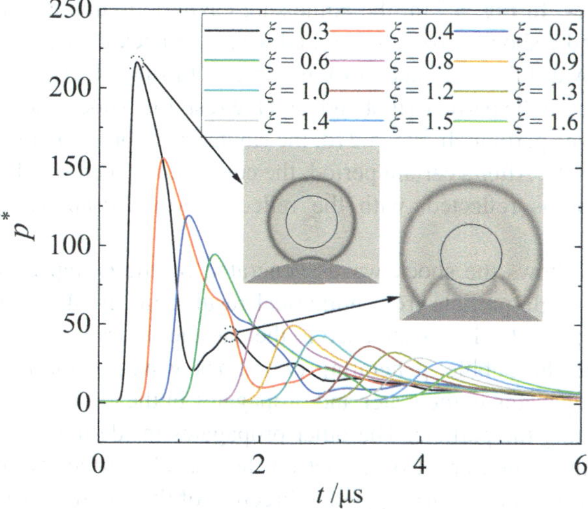

Fig. 4.20 Effect of particle to cavitation bubble distance on the variation of particle apex pressure induced by the shock wave. (Reprinted with the permission from Ref. [10] Copyright (2024) (ELSEVIER))

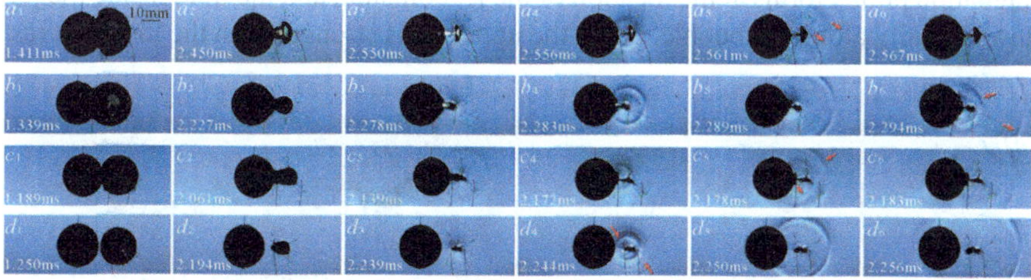

Fig. 4.21 Shock waves generated by the collapse of a cavitation bubble near a single particle as obtained from experimental measurements. (Reprinted with the permission from Ref. [11] Copyright (2023) (ELSEVIER))

ξ increases, both the durations of the pressure rise and the pressure drop grows. For small dimensionless distances, e.g., $\xi = 0.3$, the shock wave together with its reflected waves results in multiple peak pressures, with the peak pressure decreasing with each successive reflection.

4.3.2 Collapse

Figure 4.21 demonstrates the shock waves generated by the collapse of a cavitation bubble near a single particle as obtained from experimental measurements. From Figs. 4.21a–d, the distance between the particle and the spark-induced cavitation bubble increases. In Fig. 4.21a, the collapsing cavitation bubble exhibits a mushroom shape, and a shock wave is generated when the neck contracts along the particle surface toward the right apex of the particle. In Figs. 4.21b, c, the cavitation bubble collapses at the particle right end, and the resulting shock waves are reflected upon reaching the particle. In Fig. 4.21d, the cavitation bubble collapses on the right side of the particle. After a certain period, the collapse-induced shock wave impacts the particle and is reflected, with the reflected wave disappearing quickly in the water.

Figure 4.22 shows the shock waves generated by the collapse of a cavitation bubble near a single particle from numerical results. In Fig. 4.22a, the cavitation bubble collapses while closely attached to the particle. In frames 1 to 2, the jet first pierces through the bubble, generating two shock waves at the pierce site. One shock wave propagates in the water and propagates in the opposite direction of the jet after impacting the particle. The other propagates inside the cavitation bubble. During the continuous impact of the jet on the particle surface (frames 3–4), the shock waves propagating in the opposite direction of the jet are generated.

In Fig. 4.22b, the cavitation bubble collapses above the particle. In frame 1, the jet from the upper side of the bubble is strong and generates a shock wave inside the bubble that propagates toward the particle. In frame 2, two counter-directed jets collide and form shock waves that propagate both in water and inside the bubble. A stagnation points forms at the location of the jet collision, and the fluid inside the jet flows radially along the bubble. Since the upper jet is stronger than the lower jet (as

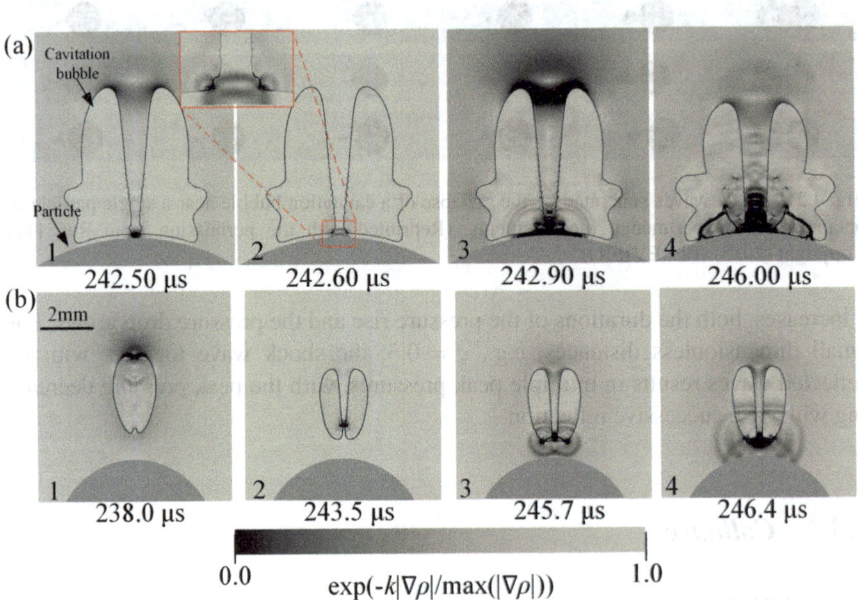

Fig. 4.22 Shock waves generated by the collapse of a cavitation bubble near a single particle as obtained from numerical results. Reprinted with the permission from Ref. [10] Copyright (2024) (ELSEVIER). (Reprinted with the permission from Ref. [12] Copyright (2024) (Springer))

seen in frame 7 of Fig. 4.15), the radial flow again moves toward the particle. When the radial flow pierces the bubble in frame 3, shock waves are produced again. In frame 4, shock waves reach the particle and are reflected.

4.3.3 Splitting

Figure 4.23 shows the shock waves generated by the splitting of a cavitation bubble between two particles. In frame 1, a high-pressure water region forms at the splitting location as indicated by the pressure distribution in Fig. 4.16c. The high-pressure region results in a shock wave propagating in the water and the formation of jets in the two daughter bubbles. The rapid development of the jets leads to the formation of a shock wave within each daughter bubble. When the jets pierce the cavitation bubble and impact the particles, the generation and the propagation of the shock waves are the same as those displayed in Fig. 4.22.

4.4 Cavitation and Erosion of Hydraulic Machinery

This section describes the synergistic damage to hydraulic machinery by sediment particles and cavitation bubbles. Hydraulic machinery such as impulse turbines, Francis turbines, and pumps operating in sediment-laden water environments often face the combined threat of sediment particles and cavitation bubbles, whose synergistic interaction can significantly accelerate the deterioration of critical components. On one hand, sediment particles directly impact the surfaces of flow

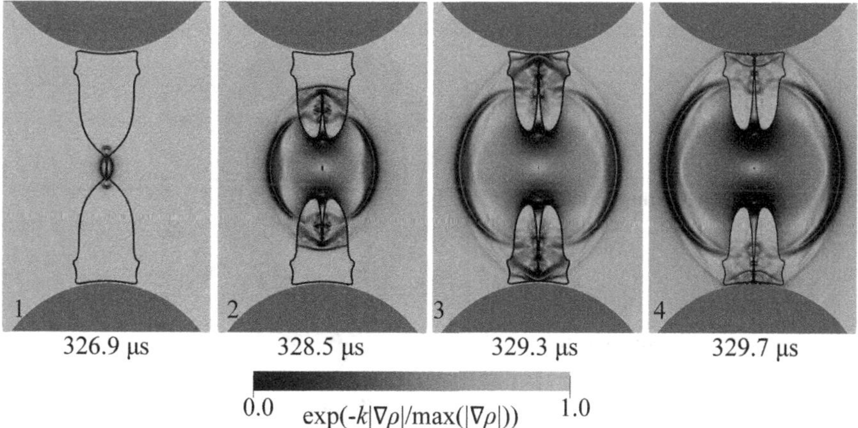

Fig. 4.23 Shock waves generated by the splitting of a cavitation bubble between two particles. (Reprinted with the permission from Ref. [7] Copyright (2023) (ELSEVIER))

components, causing wear and erosion; on the other, sediment particles create favorable conditions for the initiation of cavitation, thereby increasing both the frequency and intensity of cavitation. The collapse of cavitation bubbles releases jets and shock waves that induce pitting and fatigue damage, while further accelerating the motion of sediment particles to enhance their destructive force. The combined action of sediment particles and cavitation bubbles accelerates material erosion, drastically reducing the service life and operational efficiency of hydraulic machinery, thus presenting a critical challenge that urgently needs to be addressed in engineering practice.

4.4.1 Impulse Turbine

The Impulse turbine, also known as Pelton turbines, is widely employed in high-head, low-flow hydropower plants. Its main components include the distributor, the injector (consisting of the nozzle and needle), the runner, the casing, and the generator. Figure 4.24 illustrates the key overflow components. The distributor directs the water flow into the injector, which converts the pressure potential energy into the kinetic energy of the water jet. This water jet then strikes the runner's buckets in the tangential direction, causing the runner to rotate and drive the generator to produce electricity. The nozzle, needle and buckets are the critical components that are subjected to synergistic damage by particles and cavitation bubbles.

Figure 4.25 illustrates the abrasion mechanism in different regions of the nozzle and the needle. Under high-speed flow, nozzle region I and needle region II suffer severe abrasion with plowing and craters. The particles impacting nozzle region II and needle region I at low impact angles and low velocities produce ripples and

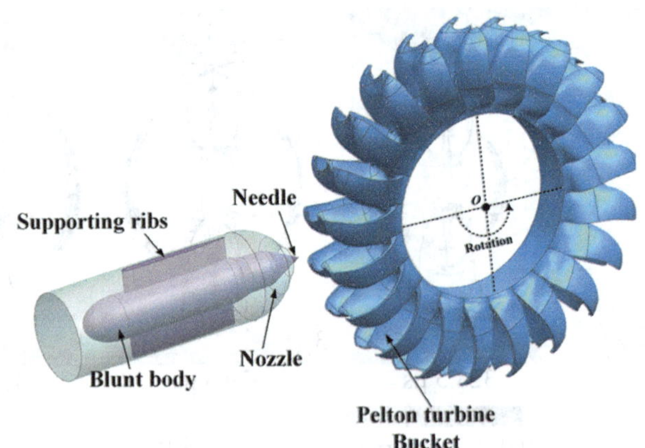

Fig. 4.24 Key overflow components of Pelton turbines. (Reprinted with the permission from Ref. [13] Open access under a CC BY 4.0 license, https://creativecommons.org/licenses/by/4.0/)

Fig. 4.25 Abrasion pattern of the nozzle and the needle. (Reprinted with the permission from Ref. [14] Copyright (2020) (ELSEVIER))

scratches [14]. Large particles, subject to stronger centrifugal and inertial forces, causes the abrasion region to be concentrated in needle region II. Small particles entrained by the flow cause abrasion concentrated in needle region I. Guo et al. [15] observed that secondary flows induced by vortex near the needle accelerate particles, intensifying needle abrasion, and reported asymmetric abrasion in the injector with greater asymmetry at the needle than at the nozzle. When the nozzle opening decreases Bajracharya et al. [16] found increases in the number of particles striking both the nozzle and the needle and in particles rebounding from the nozzle onto the needle surface, causing a sharp rise in needle abrasion rate with a slight increase in nozzle abrasion rate. Shrivastava et al. [17] pointed out that cavitation near the nozzle seat coupled with particles further aggravates damage to the injector.

For the bucket, the particles hit the splitting edge with the jet at a large angle, resulting in plastic deformation of the splitter, with the top showing a wavy shape, increasing width and decreasing height. The particles moving along the inner surface of the bucket at a small angle produce linear wear marks [18]. Small particles tend to wear the outlet edge near the bucket root while large particles preferentially wear the inlet near the splitter. An increase in flow velocity expands the wear zone on the bucket whereas changes in nozzle opening have minimal effect on the wear extent.

Figure 4.26 illustrates the abrasion mechanism and distribution pattern of the bucket during its rotation. During the feed stage (a–b), the most serious abrasion occurs due to the large number of high-speed sediment particles impinging on the cutout (E1 region). During the maintain stage (b–d), the splitter is subjected to the concentrated impingement of high-speed particles, which results in significant abrasion (E2 region). During the transition between feed and maintain, the bucket root bending region is highly abrasive due to the increased frequency of particle impacts (region E3). During the evacuate stage (d–e), the curved region at the bucket bottom is subjected to large-angle impacts of particles leading to increased abrasion (region E4).

Figure 4.27 illustrates the cavitation phenomenon that occurs around the bucket as it rotates. During the energy transfer from the jet to the bucket, the local pressure at the back of the cutout often drops below the saturated vapor pressure. As a result,

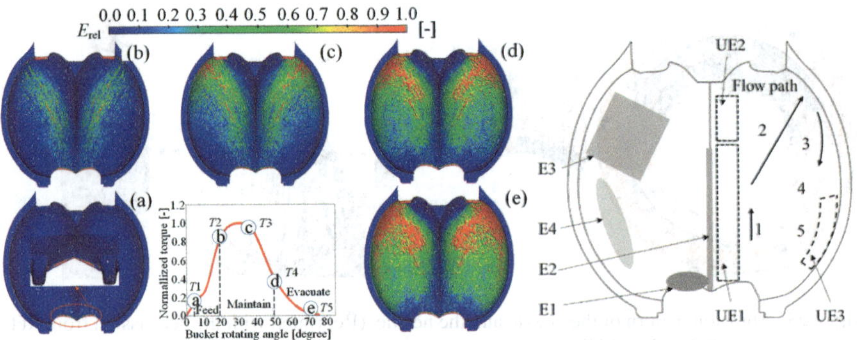

Fig. 4.26 Abrasion mechanism and distribution pattern of the bucket during its rotation. (Reprinted with the permission from Ref. [19] Copyright (2022) (ELSEVIER))

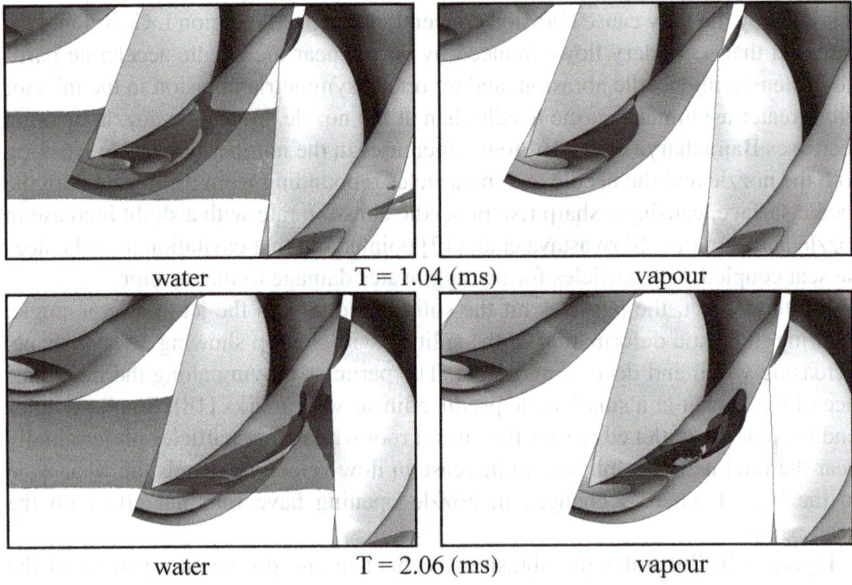

Fig. 4.27 Cavitation phenomenon that occurs around the bucket during its rotation. (Reprinted with the permission from Ref. [20] Copyright (2014) (ASME))

the back of the cutout is a high-risk area for cavitation pitting. Han et al. [21] found that under low sediment concentration conditions, cavitation reduces the number of sediment particles impacted at large angles and reduces abrasion. In contrast at high sand concentrations, cavitation bubbles increase the number of sediment particles impacted at large angles and exacerbate the abrasion phenomenon.

4.4.2 *Francis Turbine*

Francis turbine is a reaction turbine widely used in hydroelectric power stations with medium to high head (20–700 meters). Its main components include the worm shell, guide vanes, runner and tailpipe. The water enters from the worm shell, passes through the guide vanes to adjust the flow direction, and then impacts the runner blades. The runner converts the potential and kinetic energy of the water into mechanical energy, which drives the shaft to rotate and then drives the generator to produce electricity. The water flow is finally discharged through the tailpipe. The Francis turbine's runner blade design allows it to operate efficiently under different operating conditions, but its complex operating environment and flow characteristics make it susceptible to sediment abrasion and cavitation erosion.

Cavitation in hydraulic turbines arises when local pressure drops below the vapor pressure of water, leading to the formation, growth, and collapse of vapor bubbles. In Francis turbines, this phenomenon predominantly occurs at the runner outlet due to the inherent pressure decrease across the runner passage. Figure 4.28 illustrates the typical regions of cavitation erosion on the runner. The figure highlights that cavitation primarily affects the trailing edges of the blades, particularly at the junction near the hub and the shroud. This is attributed to the high flow velocities and low pressures in these zones, which facilitate the formation of cavitation bubbles.

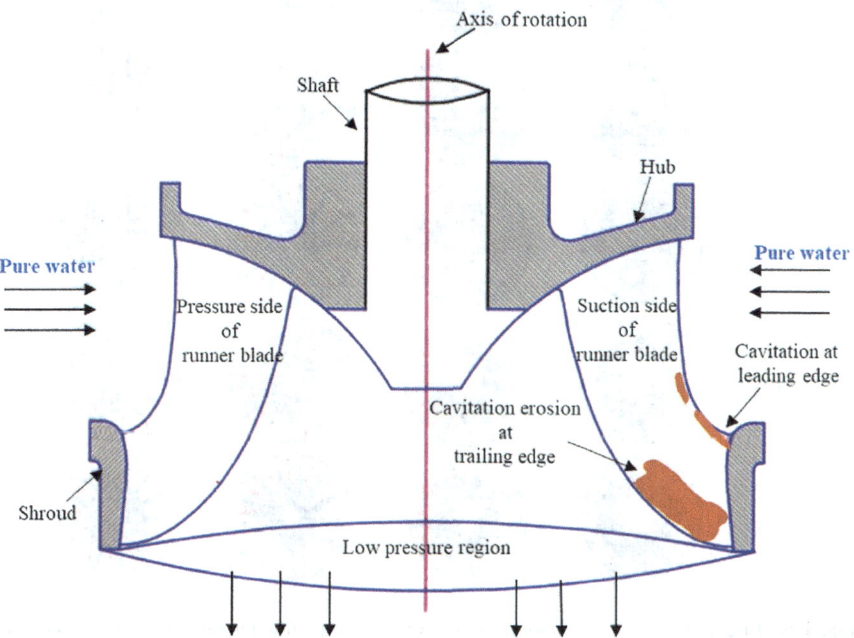

Fig. 4.28 Typical regions of cavitation erosion on the runner. (Reprinted with the permission from Ref. [22] Copyright (2024) (ELSEVIER))

Additionally, the suction side of the blades shows evidence of cavitation due to its lower pressure compared to the pressure side, making it more susceptible to bubble formation and subsequent collapse.

Figure 4.29 illustrates the effect of operating conditions on the vapor volume fraction on the runner. Partial load (e.g. 60% and 80%), full load (100%) and over-load (120%) conditions are shown. At part load, cavitation tends to concentrate in the central region of the runner. This is driven by high swirl vortexes at the runner outlet, as observed in numerical simulations. Conversely, under overload

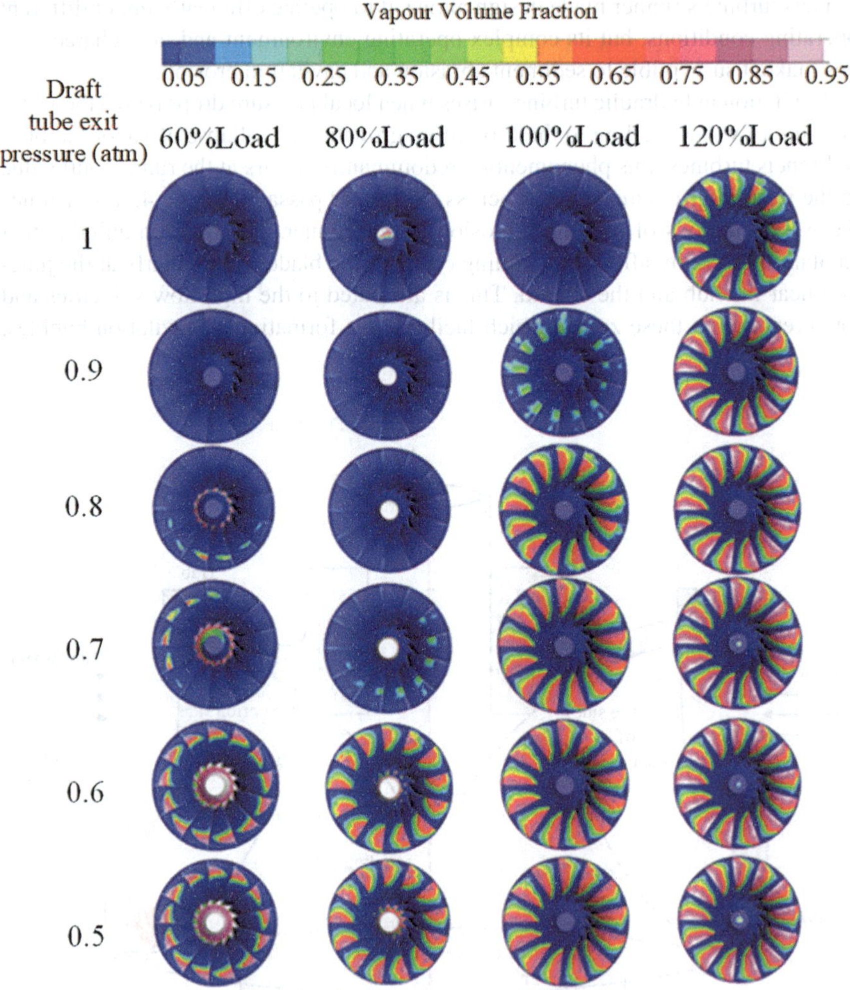

Fig. 4.29 Effect of operating conditions on the vapor volume fraction on the runner. (Reprinted with the permission from Ref. [23] Copyright (2020) (ELSEVIER))

conditions, cavitation extends to both the pressure and suction sides of the blades. This shift is due to the expansion of high-velocity, low-pressure zones as the turbine operates beyond its rated capacity. When the pressure in the draft tube decreases, the cavitation zone increases from the middle of the runner in partial and full load cases.

Sediment erosion, caused by the abrasive action of sand particles entrained in the water flow, affects different regions of the Francis turbine compared to cavitation Fig. 4.30 provides a schematic of typical areas of sediment erosion on the runner blades. It can be noticed that sediment erosion predominantly occurs at the inlet of the runner blades and on the guide vanes. These locations experience high flow velocities and large impact angles of sediment particles, resulting in significant cutting and impingement damage. The pressure side of the blades is also affected, as it is directly impacted by the flow and sediment mixture, exacerbating wear in these zones.

Figure 4.31 illustrates the effect of sediment concentration on erosion distribution. At low sediment concentrations, erosion is primarily confined to the inlet regions of the runner and the guide vanes, consistent with the high-velocity impact zones identified earlier. However, as sediment concentration increases, the erosion pattern becomes more widespread, extending to the outlet edges and suction side of the blades. This broader distribution reflects the intensified abrasive action of a higher particle load, which amplifies both the range and severity of sediment-induced wear. These observations highlight that sediment erosion locations are governed not only by flow dynamics but also by the sediment content in the water, with

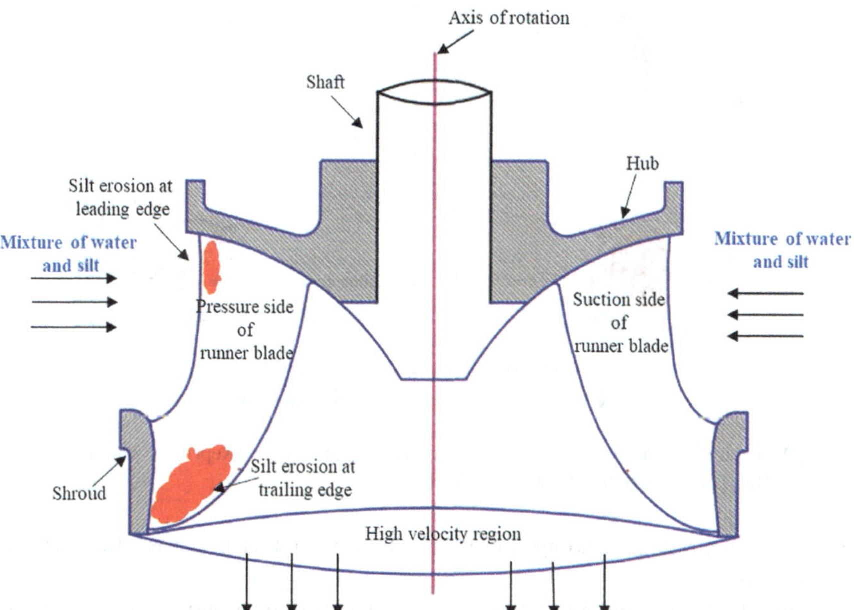

Fig. 4.30 Typical regions of sediment erosion on the runner. (Reprinted with the permission from Ref. [22] Copyright (2024) (ELSEVIER))

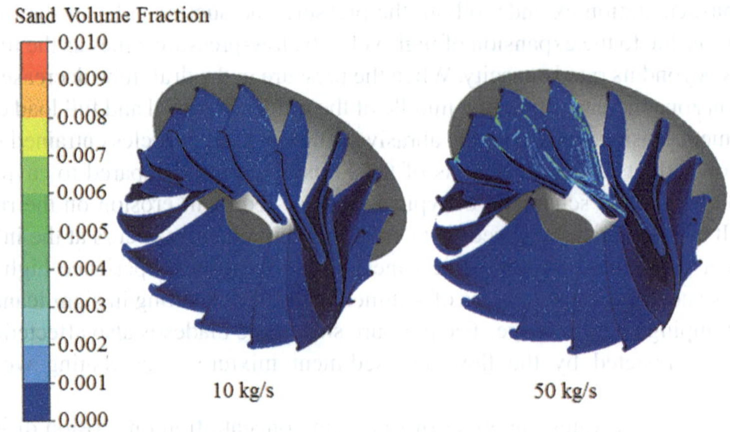

Fig. 4.31 Effect of sediment concentration on distribution of sediment erosion. (Reprinted with the permission from Ref. [24] Open access under a CC BY 4.0 license, https://creativecommons.org/licenses/by/4.0/)

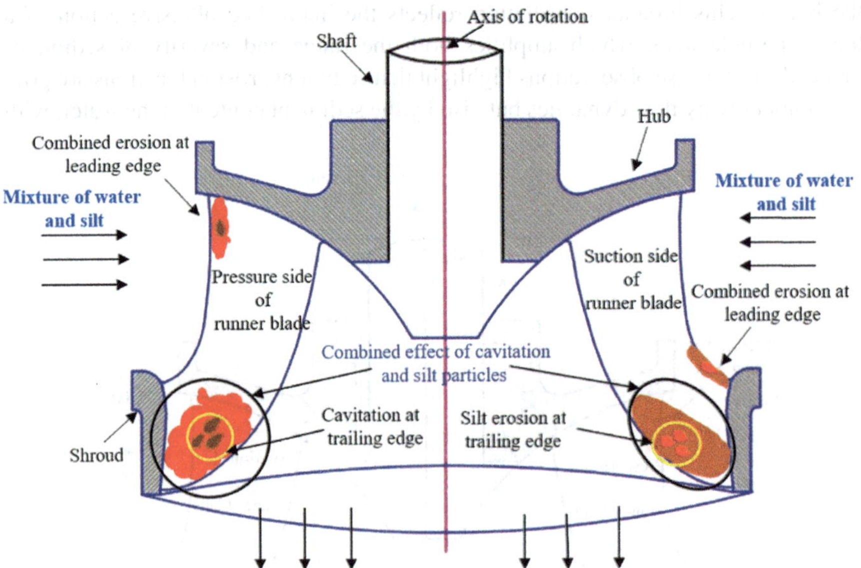

Fig. 4.32 Region of combined sediment and cavitation on the runner. (Reprinted with the permission from Ref. [22] Copyright (2024) (ELSEVIER))

higher concentrations leading to more extensive damage across the turbine components.

Figure 4.32 illustrates the region of combined sediment and cavitation on the runner. The combination of cavitation and sediment erosion severely damages the runner blades, especially at the outlet and suction side. This synergy accelerates material loss and fatigue, altering the blade geometry and degrading hydraulic efficiency. Such damage increases vibration, noise, and operational instability, ultimately shortening the turbine's lifespan, especially in sediment-rich environments. Addressing these challenges requires targeted design strategies, such as reinforcing vulnerable regions and optimizing flow paths to minimize pressure drops and sediment impact angles, ensuring enhanced durability and performance of Francis turbines under diverse operating conditions.

4.4.3 Centrifugal Pump

Figure 4.33 shows a transparent model of a centrifugal pump. The fluid enters the pump through the volute inlet, is directed by the guide vanes into the impeller. Under the action of impeller rotation, the liquid acquires both kinetic and pressure energy before being discharged through the draft tube. Due to the high-speed rotation of the impeller, a negative pressure zone is formed at its inlet, which makes the fluid continuously sucked in and realizes the continuous transportation of the liquid.

Figure 4.34 illustrates the development of cavitation bubbles within the centrifugal pump. In Fig. 4.34a ($t = 0$ ms), as fluid passes through the local low-pressure region on the blade suction side, the pressure rapidly falls below the saturated vapor pressure and microbubbles nucleate on the blade surface or at wall defects. In Figs. 4.34b–d ($t = 2$–6 ms), these bubbles grow rapidly by absorbing vapor and coalescing under the sustained low-pressure conditions. In Figs. 4.34e–g

Fig. 4.33 Transparent model of a centrifugal pump. (Reprinted with the permission from Ref. [25] Open access under a CC BY 4.0 license, https://creativecommons.org/licenses/by/4.0/)

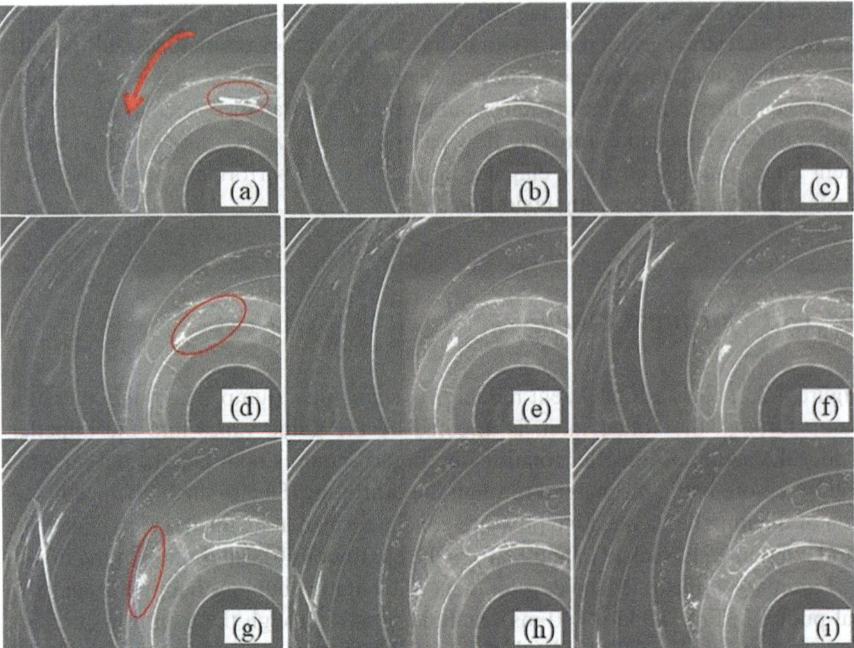

Fig. 4.34 Evolution of cavitation bubbles within the centrifugal pump. (Reprinted with the permission from Ref. [25] Open access under a CC BY 4.0 license, https://creativecommons.org/licenses/by/4.0/)

($t = 8$–12 ms), the bubbles reach a quasi-steady state, forming the cavitation cloud. Finally, In Figs. 4.34h–i ($t = 14$–16), shear forces and turbulent fluctuations fragment the cloud into numerous small-scale bubbles, generating numerous small-scale bubbles. The subsequent collapse of bubbles produces microjets and pressure waves that impinge on the pump casing surface, contributing to erosion.

Sand particles in the stream affect cavitation phenomenon in the flow channel of centrifugal pumps. The numerical results of Wang et al. [26] show that increasing the sand concentration the more pronounced the cavitation phenomenon. As the sand particle diameter increases, the volume of cavitation decreases. Figure 4.35 illustrates the distribution of blade erosion due to sand and cavitation. *NPSH*a stands for Nozzle suction head allowance. Sand erosion is distributed over all surfaces of the blade, with the most severe erosion occurring at the outlet of the pressure surface. Cavitation erosion is mainly distributed upstream of the suction surface of the blade. Comparing the erosion distribution from sand and cavitation, it can be found that the scope of sand erosion is large and the damage is small, while the scope of cavitation erosion is small and the damage is large. Under the synergistic effect of sand and cavitation, the most serious location of centrifugal pump damage is the leading edge and the upstream of the blade suction surface.

Figure 4.36 shows the influence of sand concentration on blade erosion. Cavitation erosion is greater than sand erosion. As sand concentration increases, the

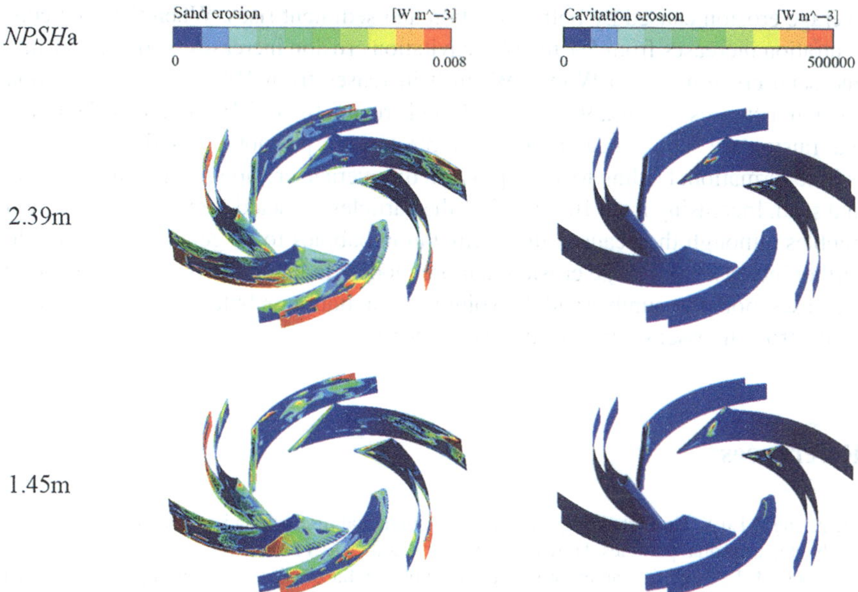

Fig. 4.35 Distribution of blade erosion due to sand and cavitation. (Reprinted with the permission from Ref. [26] Copyright (2025) (Emerald Publishing))

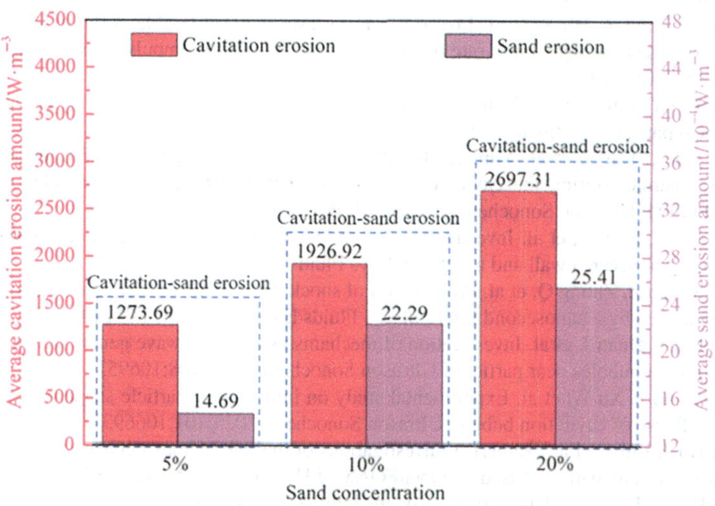

Fig. 4.36 Influence of sand concentration on blade erosion. (Reprinted with the permission from Ref. [26] Copyright (2025) (Emerald Publishing))

average erosion caused by both cavitation and sediment rises. When the sand concentration increases from 5% to 10%, cavitation erosion increases by 653.23 $W{\cdot}m^{-3}$ and sand erosion by 7.6 $W{\cdot}m^{-3}$. When it increases from 10% to 20%, cavitation erosion increases by 770.39 $W{\cdot}m^{-3}$ and sand erosion by 3.12 $W{\cdot}m^{-3}$. It can be found that raising the sand concentration from 5% to 10% promotes abundant cavitation bubble formation, leading to a sharp rise in the average erosion from both cavitation and sand. Increasing from 10% to 20%, the particles attracted some of the cavitation bubbles although they caused many cavitation bubbles to be generated, leading to further increase in average erosion is more modes. In addition, as particle diameter increases, both the number and the volume of cavitation bubbles decrease, the cavitation erosion lessens while sand erosion increases.

References

1. Zhang Y, Lu X, Hu J, et al. Experimental and numerical research on jet dynamics of cavitation bubble near dual particles. Ultrason Sonochem. 2025;112:107168.
2. Koch M, Lechner C, Lauterborn W, et al. Bubble collapse directly at an object: fast jet and shock wave. Proc Meetings Acoustics. 2022;48:032001.
3. Zhang Y, Chen F, Zhang Y, et al. Experimental investigations of interactions between a laser-induced cavitation bubble and a spherical particle. Exp Thermal Fluid Sci. 2018;98:645–61.
4. Zheng X, Wang X, Lu X, et al. Experimental research on the collapse dynamics of the cavitation bubble near two spherical particles. J Mech Sci Technol. 2023;37(5):2451–60.
5. Zhang Y, Ding Z, Hu S, et al. Investigation on laser-induced bubble collapse among triple particles based on high-frame-rate photography and the Kelvin impulse model. Phys Fluids. 2024;36(5):053304.
6. Hu J, Liu Y, Liu Y, et al. Numerical investigation of cavitation bubble jet dynamics near a spherical particle. Symmetry. 2023;15(9):1655.
7. Hu J, Lu X, Liu Y, et al. Numerical and experimental investigations on the jet and shock wave dynamics during the cavitation bubble collapsing near spherical particles based on OpenFOAM. Ultrason Sonochem. 2023;99:106576.
8. Hu J, Liu Y, Duan J, et al. Investigations on the jets and shock waves of a cavitation bubble collapsing between a wall and a particle. Phys Fluids. 2024;36(3):033330.
9. Geng S, Yao Z, Zhong Q, et al. Propagation of shock wave at the cavitation bubble expansion stage induced by a nanosecond laser pulse. J Fluids Eng. 2021;143(5):051209.
10. Hu J, Liu Y, Duan J, et al. Investigation of mechanisms of shock wave generation by collapse of cavitation bubbles near particles. Ultrason Sonochem. 2024;108:106952.
11. Zou L, Luo J, Xu W, et al. Experimental study on influence of particle shape on shockwave from collapse of cavitation bubble. Ultrason Sonochem. 2023;101:106693.
12. Yu J, Hu J, Liu Y, et al. Numerical investigations of the interactions between bubble induced shock waves and particle based on OpenFOAM. J Hydrodyn. 2024;36(2):355–62.
13. Zhao H, Zhu B, Xu, et al. Investigation on the influence of bucket's flow patterns on energy conversion characteristics of Pelton turbine. Eng Appl Comput Fluid Mech. 2023;17(1):2234435.
14. Din M, Harmain G. Assessment of erosive wear of Pelton turbine injector: nozzle and spear combination—a study of Chenani hydro-power plant. Eng Fail Anal. 2020;116:104695.
15. Guo B, Xiao Y, Rai A, et al. Sediment-laden flow and erosion modeling in a Pelton turbine injector. Renew Energy. 2020;162:30–42.
16. Bajracharya T, Shrestha R, Sapkota A, et al. Modelling of Hydroabrasive erosion in Pelton turbine injector. Int J Rotating Mach. 2022;2022(1):9772362–15.

17. Shrivastava N, Rai A, Abbas A, et al. Analysis of hydro-abrasive erosion in a high-head Pelton turbine injector using a Eulerian-Lagrangian approach. Proc Inst Mech Eng Part A J Power Energy. 2024;238(3):495–514.
18. Padhy M, Saini R. Study of silt erosion mechanism in Pelton turbine buckets. Energy. 2012;39(1):286–93.
19. Xiao Y, Guo B, Rai A, et al. Analysis of hydro-abrasive erosion in Pelton buckets using a Eulerian-Lagrangian approach. Renew Energy. 2022;197:472–85.
20. Rossetti A, Pavesi G, Ardizzon G, et al. Numerical analyses of Cavitating flow in a Pelton turbine. J Fluids Eng. 2014;136(8):081304.
21. Han L, Guo C, Yuan Y, et al. Investigation on sediment erosion in bucket region of Pelton turbine considering cavitation. Phys Fluids. 2024;36(2):023305.
22. Kumar P, Singal S, Gohil P. A technical review on combined effect of cavitation and silt erosion on Francis turbine. Renew Sust Energ Rev. 2024;190:114096.
23. Tiwari G, Kumar J, Prasad V, et al. Derivation of cavitation characteristics of a 3MW prototype Francis turbine through numerical hydrodynamic analysis. Mater Today Proc. 2020;26:1439–48.
24. Rakibuzzaman M, Kim H, Kim K, et al. Numerical study of sediment erosion analysis in Francis turbine. Sustainability. 2019;11(5):1423.
25. Cui Y, Cheng B, Ding Q, et al. Study on cavitation bubble characteristics in centrifugal pump based on image recognition. PRO. 2023;11(12):3314.
26. Wang X, Wang Y, Chen J, et al. Numerical investigation on cavitating flow and cavitation-sand erosion characteristics of a centrifugal pump under sand-laden water conditions. Eng Comput. 2025;42(3):1163–85.

Chapter 5
Conclusion

This book offers a comprehensive and detailed examination of multiphase flow with bubbles. It explores the dynamics of bubbles in various multiphase systems, addressing their behavior under acoustic excitation, the propagation of sound waves in vapor/gas/liquid mixtures, and cavitation phenomena in vapor/liquid/solid environments. Conclusions are given as follows.

Firstly, the bubble dynamics under acoustic excitation is introduced. It began by deriving first-and second-order equations of motion for a single bubble, highlighting their differences in predicting oscillatory behavior. The mass transfer mechanisms at the bubble interface were examined, and threshold conditions for bubble growth and dissolution under both single-and dual-frequency acoustic fields were identified. Furthermore, the energy dissipation mechanisms, encompassing viscous, thermal, and acoustic damping are detailed analyzed. The analysis of dual-frequency excitation revealed resonance behaviors, scattering effects, and secondary Bjerknes forces.

Secondly, the propagation of sound waves in vapor/gas/liquid multiphase systems is introduced. Various wave speed prediction models are derived and compared across different frequency regimes. A detailed frequency-partitioning framework to distinguish low-, mid-, and high-frequency wave behaviors is presented. Key parameters such as void fraction, vapor mass fraction, and bubble radius were shown to significantly influence critical frequency thresholds, minimum wave speeds, and damping characteristics.

Finally, cavitation in vapor/liquid/solid multiphase flows is investigated. The behavior of cavitation bubbles, including the collapse morphology and splitting behavior near single and multiple particles, as well as the effects of key parameters are presented. The formation mechanisms and evolutionary properties of jets are analyzed in detail. The number and position of particles affect the jet characteristics.

J. Hu et al., *Multiphase Flow with Bubbles*, SpringerBriefs in Energy, https://doi.org/10.1007/978-3-031-99216-2_5

At the inception of cavitation bubble, a shock wave is formed by the release of high pressure inside the cavitation bubble. During cavitation bubble collapse, shock waves are generated when the cavitation bubble splits, the jet penetrates the bubble wall, and the jet impacts the bubble wall or particles. The synergistic erosive effects of cavitation bubbles and sediment particles on the surfaces of over-flow components are emphasized with Impulse turbines, Francis turbines and centrifugal pumps as representatives of hydraulic machinery.

Index